Horticultural Crops

Emerging Issues and Techniques

The Editors

Dr. Shailendra Kumar Pandey, Prof and Head, Department of Horticulture, College of Agriculture, JNKVV, Jabalpur, Madhya Pradesh has 35 years of experience in teaching, research and extension. His area of specialization is Nursery Management and Commercial Fruit Production. He has guided 07 PhD and 35 MSc students as major guide. He has handled several project as Principle Investigator funded by BARC, ICAR, NHB-GoI and RKVY. He has published more than 54 Research Paper, 02 Books/ Practical manual, 02 Book chapters, 02 Technical bulletins and more than 150 popular articles. He has received Life Time Achievement award by Astha Foundation, Meerut in 2018.

Dr Reena Nair, Assistant Professor, Department of Horticulture, College of Agriculture, JNKVV, Jabalpur, Madhya Pradesh has 10 years of teaching and research experience. Her area of specialization is Vegetable Science (Mutation Breeding). She had a brilliant academic records throughout and has qualified National Eligibility Test (NET) in 2009. She has guided 09 MSc Vegetable Science Students and 01 PhD Vegetable Science student as major advisor. Apart from this she has also guided 11 students as members of their Advisory Committee. She is teaching courses of Horticulture in BSc, MSc and PhD classes. She has published research papers 2 in International Journal and 30 in National Journals, 18 Abstracts, 04 Book Chapters, 05 Books/ Manuals/ Monographs, 22 popular articles and 01 Proceeding/Abstract Book.

Dr Kanchan S. Bhan, Scientist, Department of Horticulture, College of Agriculture, JNKVV, Jabalpur, Madhya Pradesh, received her PhD degree from Rani Durgawati University, Jabalpur and joined as Post-Doctoral Research Associate in Cellular and Molecular Biology Program at the University of Florida. Later she started working as Assistant Professor in the Department of Plant Molecular Biology and Biotechnology at Indira Gandhi Krishi Vishwavidyalaya, Raipur almost four years in teaching and research activities. Currently, she is working as a Vegetable Breeder in All India Coordinated Programme at JNKVV, Jabalpur. Additionally she worked in different national and international projects and awarded fellowships. She has guided several MSc and PhD students, and published more than fifteen research papers in National and International peer reviewed journals.

Horticultural Crops

Emerging Issues and Techniques

The Editors

Dr. Shailendra Kumar Pandey, Prof and Head, Department of Horticulture, College of Agriculture, JNKVV, Jabalpur, Madhya Pradesh has 35 years of experience in teaching, research and extension. His area of specialization is Nursery Management and Commercial Fruit Production. He has guided 07 PhD and 35 MSc students as major guide. He has handled several project as Principle Investigator funded by BARC, ICAR, NHB-GoI and RKVY. He has published more than 54 Research Paper, 02 Books/ Practical manual, 02 Book chapters, 02 Technical bulletins and more than 150 popular articles. He has received Life Time Achievement award by Astha Foundation, Meerut in 2018.

Dr Reena Nair, Assistant Professor, Department of Horticulture, College of Agriculture, JNKVV, Jabalpur, Madhya Pradesh has 10 years of teaching and research experience. Her area of specialization is Vegetable Science (Mutation Breeding). She had a brilliant academic records throughout and has qualified National Eligibility Test (NET) in 2009. She has guided 09 MSc Vegetable Science Students and 01 PhD Vegetable Science student as major advisor. Apart from this she has also guided 11 students as members of their Advisory Committee. She is teaching courses of Horticulture in BSc, MSc and PhD classes. She has published research papers 2 in International Journal and 30 in National Journals, 18 Abstracts, 04 Book Chapters, 05 Books/ Manuals/ Monographs, 22 popular articles and 01 Proceeding/Abstract Book.

Dr Kanchan S. Bhan, Scientist, Department of Horticulture, College of Agriculture, JNKVV, Jabalpur, Madhya Pradesh, received her PhD degree from Rani Durgawati University, Jabalpur and joined as Post-Doctoral Research Associate in Cellular and Molecular Biology Program at the University of Florida. Later she started working as Assistant Professor in the Department of Plant Molecular Biology and Biotechnology at Indira Gandhi Krishi Vishwavidyalaya, Raipur almost four years in teaching and research activities. Currently, she is working as a Vegetable Breeder in All India Coordinated Programme at JNKVV, Jabalpur. Additionally she worked in different national and international projects and awarded fellowships. She has guided several MSc and PhD students, and published more than fifteen research papers in National and International peer reviewed journals.

Horticultural Crops

Emerging Issues and Techniques

— Editors —

Dr. S.K. Pandey
University Professor and Head

Dr. Reena Nair
Assistant Professor

Dr. Kanchan S. Bhan
Scientist

Department of Horticulture,
Jawaharlal Nehru Krishi Vishwa Vidyalaya
Jabalpur – 482 004, Madhya Pradesh

BIOTECH

2022

BIOTECH BOOKS®

ISBN 978-81-7622-505-2

Published by: **BIOTECH BOOKS®**
4762-63/23, Ansari Road, Darya Ganj,
New Delhi - 110 002
Phone: 09899295259, 011-43003222
E-mail: biotechbooks@yahoo.co.in
Website: www.biotechbooks.in

Typeset at: **Classic Computer Services**
Delhi - 110 035

Printed at: **Replika Press Pvt. Ltd.**

PRINTED IN INDIA

Preface

Horticulture contributes to quality of life, and the beauty, sustainability, rehabilitation of our environment and the human condition. It is a boon of nature which is refined by human skills as science to obtain more and more benefits. It includes the improvement, production, distribution and use of vegetables, fruits, woody landscape, medicinal and aromatic plants, spices and condiments, plantation crops, greenhouse plants and processing. It is one of the fastest growing industries with striking professional opportunities. Horticulture production in India has more than doubled approximately from 146 million tonnes in 2001-02 to 314 million tonnes in 2018-19. India is second largest producer of fruits and vegetables in the world. Area under horticulture was also projected to rise 2.7 per cent during 2021, against the previous year. Total horticulture production in financial year 2021 is estimated at 326.58 mt, an increase of 5.81 mt over 2020. People are becoming more health conscious and the promotion of healthy habit benefits by utilizing fresh, minimal processed, dried fruits and vegetables as well as their juices and extracts is on the rise.

The sustainability of horticultural crops depends on the use of resources as well as accountable use of objectionable compounds such as pesticides which will not compromise the quality of the environment. Therefore the demand for useful and applicable information is growing as horticulture become more sophisticated focussing on sustainable yield in the present changing climatic conditions. Due to constantly changing environmental conditions, abiotic stresses are an imminent threat to horticultural crops and plant growth regulators especially Brassinosteroids regulates and maintain various physiological activities under stress conditions and play a major role in mitigating these effects.

This book "Horticultural Crops: Emerging Issues and Techniques" offers a comprehensive resource on various aspects of Horticulture. This book contains thirteen chapters written by renowned researchers on contemporary topics *viz.*, status, challenges and export potential of low volume high value crops-spices, advances in post harvest management of citrus, Microbes as Biofertilizers - Potential Approach for Sustainable Crop Production, mitigating climate change, role of plant growth regulators Brassinosteroids in salinity stress, livelihood generation and skill development of tribals through plantation crops advances in improvement of brinjal, gene tillering, Mushrooms: A potential tool for Myco-remediation and edible flowers.

This book hopes to meet the requirement of students, researchers, educators and all those who are interested in Horticulture.

Dr. S.K. Pandey

Dr. Reena Nair

Dr. Kanchan S. Bhan

Contents

Chapter 1

Realizing the Export Potential of Spices: Challenges and Strategies

K. Nirmal Babu[1], Lijo Thomas[2] and Sharon Aravind[2]

[1]Ex Director and [2]Scientist
ICAR Indian Institute of Spices Research,
Kozhikode – 673 012, Kerala

Introduction

The fame of Indian spices, established over centuries of trade spreading across the globe through diverse trade networks, confers a definite global market advantage for the spices sourced from the country. The spices from the country are often able to command a price premium due to the superior quality attributed to the spices produced here. However, the opening up of the global trade markets and the progress made in information and communication technology across the world has thrown up new challenges before the spice economy of the country. We are going through a phase, where competition for markets for traded spices are at levels, so far not witnessed. The exclusivity or monopoly in production of some spices like turmeric and black pepper has declined significantly affecting the price setting power of the country in spice exports. Along with the increased competitiveness of global spice markets, the domestic production environment has witnessed significant changes raising significant challenges to our status as global leaders in spices export. The increased stringency in quality parameters of the exported produce and rising levels of awareness about food safety and sustainability of production methods employed has increased the levels of complexity in production of spices for global consumption.

India continues to be the largest producer, consumer and exporter of spices in the world. However, when you take a look at the global trade scenario of individual spice commodities, the level of market power has declined considerably in several spice commodities like black pepper, nutmeg, turmeric *etc.* There is no doubt that India enjoys certain natural advantages like diversity of the agro-climatic conditions, rich natural diversity in germplasm resources, significant overheads of traditional knowledge of spice farming, when it comes to production of spices. But, retaining and strengthening our position as a global leader in spice supply chain would require adroit interventions in production systems, processing and value addition.

This chapter takes a brief look at the export scenario of spices in the country to have a grasp of the status of spice export sector before presenting some of the key trends which has the potential to redefine and disrupt the existing spice export patterns. Our ability to adapt to these changes will, to a large extent, determine our ability to consolidate our past gains and retain our strategic advantage in the world spice markets. The section following this outline some of the key challenges in the production sector of spices followed by an outline of policy elements for strengthening and revitalizing the spice exports from India to the world markets.

Status and Trends in Spice Economy

The spices production and availability in the country has shown a robust growth during the recent times. The recent trends in spice economy in general and the export sector in particular has been promising in terms of growth, diversification and unit value realization. The growth in domestic production of spices also creates a need to find markets for efficient transfer of output prices to the primary producers. Several factors like the rising disposable incomes, strong economic growth, growth and diversification of food processing sector in the country has increased the domestic demand for spices. However, the rising domestic availability also necessitates the exports of the surplus domestic production to world markets at competitive prices for disposal of the produce. The increase in domestic spices availability is clear from the Table 1.1 which compares the production of spices at the turn of the century to the current production status.

Table 1.1: Change in Area and Production of Major Spices in India

Crop	*2001-02*		*2016-17*	
	Area (Ha)	*Production (T)*	*Area (Ha)*	*Production (T)*
Pepper	219380	62440	134280	57000
Cardamom (Small)	72660	11370	69357	17990
Cardamom (Large)	30010	5850	26617	5570
Chilli	880000	1069000	864730	2394320
Ginger	86200	288000	160480	1047190
Turmeric	167100	562800	248050	1215520
Coriander	433400	318700	672760	863530
Cumin	521300	176100	780920	500360
Celery	12300	10000	4010	5510
Fennel	39900	38500	89540	148560
Fenugreek	115600	136600	220670	310070

Crop	2001-02		2016-17	
	Area (Ha)	Production (T)	Area (Ha)	Production (T)
Ajwain	13880	9410	34340	27840
Garlic	115200	530800	302980	1722750
Tamarind	62000	193800	49660	202150
Clove	1900	1000	2200	1230
Nutmeg	7850	1990	22760	14020
Cinnamon	338	70	240	150
Total spices	2779018	3416430	4031700	8610810

Source: Spices Board.

In case of most of the spices, the area and production has increased substantially during this period. In case of crops like turmeric, chillies, and ginger, the production has more than doubled indicating the urgent need to find markets for absorbing the enhanced production, either domestically or at the global level.

The performance of the spices export sector mirrors the primary production sector. The export volumes surged by nearly 30 per cent during the last five years. The initial estimates from Directorate General of Foreign Trade indicate that the value of spice exports crossed three billion USD during the year 2017-18. The export value has recorded robust growth over the last few years recording a 41.5 per cent growth over the last six years (Figure 1.1). The trend line indicates an yearly growth of nearly 180 million US dollars in spice export value in the short run.

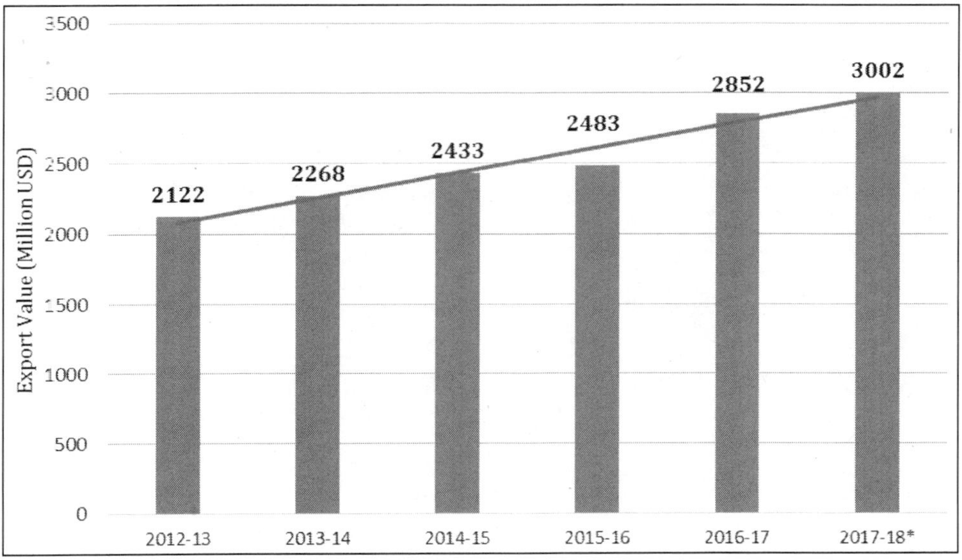

Figure 1.1: Export Value of Spices (Million USD).

***Source*: Compiled from Spices Board data and Monthly Bulletin on Foreign Trade Statistics April 2018.**

The export volume increase, both in terms of value and quantity has been broad based in the sense that the growth has occurred across several of the key spice crops in the country, which indicates the sustainable nature of the growth in export demand. The export quantity and value of selected major spice commodities is given in Table 1.2.

Table 1.2: Export Volume and Value of Selected Spice Commodities 2017-18

Spice Commodity	*Quantity ('000 Kg)*	*Value (Million INR)*
Black pepper	18340	7955
Ginger	27240	2329
Turmeric	109193	11386
Chilly	477273	45566
Large cardamom	779	586
Small cardamom	6133	6458
Nutmeg and Mace	5542	2232
Garlic	50790	3228
Seed spices		
Coriander	36664	2820
Cumin	160502	26518
Fenugreek	29509	1294
Fennel/Caraway	34871	2634

The spice commodities is gaining importance in the agricultural export basket. The share of spices, which was very less than one per cent of the total value of agricultural exports during 1980-81, showed a steady growth since and has improved its share to more than 8 per cent during 2016-17 (Table 1.3)

Table 1.3: Share of Spices in Agricultural Exports from India

Year	*Value of Spice Exports (million USD)*	*Value of Agricultural Exports (million USD)*	*Share (Per cent)*
1980-81	14	2601	0.54
1990-91	133	3521	3.78
2000-01	354	6256	5.66
2010-11	1768	24448	7.23
2016-17	2852	33816	8.43

Key Trends to Note in Spices Export

The ability to respond to changes quickly, efficiently and seamlessly will determine how the spice sector can cope with the disruptions in the production environment and market demand conditions across the global markets. Some of the trends in the spice export market are indicated below

- ☆ A significant shift from export demand for raw materials to demand for value added spice extracts and products like oleoresins and essential oils. The shift towards value added products in exports is a positive trend indicating the retaining of more stages of the value chain within the country with additional income generation and employment opportunities.
- ☆ The increasing share of non-traditional spices like mint and mint products in the export basket. The traditional export basket of spices has undergone changes and some of the crops, where India was not a major market player has gained dominance in the spice export market. The rise of mint crop cultivation is a typical example, where over a span of three decades, India has emerged as a significant supplier of mint and mint products. Similar non-traditional spices which can be adapted to India conditions need to be explored.
- ☆ The role of medicinal and pharmaceutical applications as end destination of the spice products demanded has brought in several changes in the quality and process of spice exports. The medical application of spices and the pharmaceutical preparations using spice ingredients require higher quality and need to follow stringent norms with respect to their purity. The rise of wellness industry and the increasing application of spices as functional foods has also driven this change.
- ☆ The diversification of destinations is also a major change in the spice export sector. The diversity of destinations has increased with direct trade links being opened up with hitherto unreached countries in our quest for seeking out niche markets and communities where the spices as a export commodity might be better valued and appreciated. The institutional changes in multilateral global trade facilitation and progress in transport networks have generally aided this trend.

The growth in export quantity and export value of spices has been robust in the recent past. During the last two decades the export of spices grew by 9.1 per cent in terms of export quantity with the compound annual growth rate of cumin being the highest (Table 1.4).

Table 1.4: Growth Rate of Export Quantity and Value of Spices 1998-99 to 2017-18

Crop	*Export Quantity*	*Export Value*
Pepper	-1.76	8.23
Cardamom	12.80	18.00
Turmeric	6.43	15.57
Ginger	8.55	15.86
Chili	11.73	20.11
Cumin	17.47	23.51
Coriander	5.58	14.60
Mint Products	9.20	19.22

Crop	Export Quantity	Export Value
Oil and Oleoresins	8.05	12.93
Curry Powder and Paste	10.17	17.65
Spices Total	9.10	15.68

Challenges in the Spice Production Sector

The ability of the primary production sector of spices to generate domestic surplus depends on the progress in addressing the challenges in the production sector. The major challenges and constraints in the spice crop production sector are

- Climate change resulting in drought/excess moisture, high/low temperature during critical periods, *etc.*
- Emergence and epidemics of pests and diseases.
- Adulteration of spices at multiple points in the value chain
- Pesticide residues and mycotoxin contaminants in the products and lack of MRL and ADI standards in some of the pesticides used in spices
- Shifting of interests of growers to more profitable/less risky crops.

Our capability to address these challenges directly depend on the technological options developed in response to these challenges. The developed solutions need to be simple, cost-effective and farmer-friendly. At present, the research establishment and other development institutions involved in spices sector have been able to successfully address the biotic challenges and institutional lacunae in the spices sector. But the financial support and resource allocation for the sector need to be enhanced to develop the capabilities of the public funded institutional stakeholders to continually evolve strategies to stay ahead of the emerging challenges in the spices sector.

Key Areas for Improvement and Policy Imperatives

The spices export sector, as discussed in the sections above have shown tremendous progress in the recent decades and has shown the promise to be one of the significant contributors for the agricultural development in the country. Substantial latent potential exist in the spice crops production and export marketing to sustain the robust growth and development in the sector. However, this would require some strategic policy interventions and holistic institutional support for the sector. Some of the areas for enhancing the export potential of spices and the policy elements for achieving these objectives are indicated below.

- Varietal development efforts with a focus on future global market demands and suiting the resource profile of the farming community need to be evolved. The varieties need to be climate smart and flexible across cropping systems.
- Focus on adoption of good agricultural practices in production and processing of spices with a vision of developing the country as a destination for food safe spices for the world.

- Leveraging the genetic diversity and germplasm resources for identifying and developing heirloom varieties with superior organoleptic profile and suited for specific global market demands
- Focus on research and development of novel chemical entities with high value applications across medical, pharmaceutical and wellness industry
- Enhancing system capability to ensure farm level traceability of spices which are exported through adoption of cutting edge technologies like block chain technology and other innovative traceability systems
- Standardization and rationalization of pesticide MRLs and other quality parameters for spice commodities across geographic destinations need to be taken up at the multilateral forums to facilitate spice trade
- Developing export trade intelligence system for facilitating and equipping domestic spice producers and exporters with real time demand forecast and other information feedback services to value chain entities.
- Marketing reforms need to be taken up with the specific aim of bringing in varietal specific marketing in place of bulking of the produce across different varieties and quality profiles. This will ensure better prices for the higher quality of produce
- Emerging institutional innovations in farmer collectivization like Farmer Producer Organizations need to be supported for aggregation and value addition in spice commodities and these collectives need to be empowered with the requisite capabilities for directly accessing the global markets for sale of the spice commodities
- Focused and result oriented public private partnership in addressing the specific knowledge gaps and research gaps in spice commodities, crop production constraints and product development bottlenecks need to be taken up
- Infrastructural development in areas like cold storage chains, terminal market linkages and alternate transportation facilities need to be creatively planned and developed to link production centres with the trading regions and terminal markets to facilitate smooth flow of goods and supporting services.

The spices sector in the country is undergoing a rapid transformation with game changing innovations and improvements in the production technology, postharvest processing, value addition and value chains. The spices exports from the country has grown from strength to strength in the recent decades, but the latent potential in the sector need to be harnessed for ensuring enhanced livelihood security of the primary spice producers and other stake holders along the spice value chains. Nurturing and improving sound techniques, technologies and innovations in the entire spices sector can help in surmounting the challenges posed by competing countries and help in meeting the global market demands while ensuring a sustainable and equitable production model.

Chapter 2

Advances and Challenges in Citrus Production and Postharvest Management

M.S. Ladaniya[1] and Dinesh Kumar[2]

[1]Ex-Director and [2]Principal Scientist (Hort.)
ICAR-Central Citrus Research Institute, Nagpur, Maharashtra

Introduction

Although citrus is grown in 114 countries, the commercial cultivation takes place in 45 countries only with a total production of 139.8 million tonnes (FAO Statistics, 2017). On production basis China tops the list with 35.47 million tonnes followed by Brazil with 19.07 million tonnes and USA is fourth with 8.54 million tonnes. India with 11.15 million tonnes is at 3rd position. Mexico with 7.82 million tonnes and Spain with 7.06 million tonnes are at 5th and 6th position, respectively (2017). The world citrus is dominated by sweet orange with 56.86 per cent contribution followed by mandarins with 25.72 per cent, limes and lemons 10.86 per cent and rest of the 6.56 per cent contributed by grapefruit and other citrus fruits. In India in the year 2017-18, the area under citrus is 10.06 lakh ha with a production of 125.10 lakh tonnes. Total mandarin production in India is 49.82 lakh tons with 4.30 lakh ha area while limes/lemons have occupied an area of 2.36 lakh ha, with 25.06 lakh tonnes production. Sweet oranges have occupied 1.83 lakh ha area with 33.69 lakh tonnes production (NHB, 2017-18). During 1961 to 2017-18 area expansion has been recorded by about 1100 per cent and production increase by 1500 per cent.

Indian citrus industry has certain unique features and unlike China and Japan where citrus is grown mainly under subtropical climate. The major citrus producing states include Andhra Pradesh, Telangana, Maharashtra, Madhya Pradesh and Punjab with total production of 18.28, 17.25, 16.61, 13.32 and 11.68 lakh tonnes respectively (NHB, 2015-16). The most important commercial citrus cultivars in India are the mandarins *i.e.*, loose jacket oranges (*Citrus reticulata* Blanco), followed by sweet orange *i.e.* tight jacket orange (*Citrus sinensis* Osbeck) and acid lime and lemons (*Citrus aurantifolia* Swingle) sharing 39.82 per cent, 26.93 per cent and 20.03 per cent respectively of all citrus fruits produced in the country (NHB, 2017-18). The indigenous species are commonly grown in homesteads and the introduced species/cultivars, which have adapted in different geographical regions, are the basis of commercial plantations.

India though is the third largest citrus producer contributing 11.15 per cent the world's total citrus production but has no place in world trade because of a number of reasons. The major being absence of seedless varieties of mandarin and low production of exportable quality fruit. Colour development of fruit is not satisfactory in tropical climate. The future strategies, therefore, must address to these problems.

Table 2.1: Production of Citrus Fruits in Major Citrus growing Countries in the World (2017)

Sl.No.	*Country*	*Production (million tonnes)*
1.	China	35.47
2.	Brazil	19.07
3.	India	11.15
4.	USA	8.54
5.	Mexico	7.82
6.	Spain	7.06

Source: Fresh Citrus, Annual Statistics, FAO 2017.

Growth of Citrus Industry in India

Between 1961 and 2017-18, area under citrus fruits in India increased from 90.7 thousand ha to 1009 thousand ha, thus registering more than ten fold increase during that period. The average annual growth rate in production during 1961 to 2012 was 20.3 per cent in terms of area and 20.9 per cent in terms of production. Mandarins, sweet oranges, limes and lemons are major citrus fruits in the country. Grapefruits and pummelos are limited in cultivation.

Table 2.2: Area and Production of Citrus during 1961 to 2017-18

Year	*Area (000, ha)*	*Production (000, tons)*	*Source*
1961	90.7	823.7	Directorate of Economics and Statistics, Ministry of Food and Agriculture, Govt. Of India
1968	105.4	1211.9	India's Export Potential for Fresh and Processed Fruits and Vegetables, 1968
1988	279.4	2282.2	Area and Production of Fruits and Vegetables in India- States Profile, NHB, New Delhi
1992	386.9	2820.0	Area and Production of Fruits and Vegetables in India – NHB, New Delhi

Year	Area (000, ha)	Production (000, tons)	Source
2004	563.0	5680.0	Statistic of Indian Horticulture Database, ICAR, New Delhi
2007	798.0	7145.0	Area and Production of Fruits and Vegetables in India-NHB, New Delhi (2007-08)
2009	923.0	8608.0	Area and Production of Fruits and Vegetables in India- NHB, New Delhi (2008-09)
2012-13	1042.5	10089.7	Indian Horticulture Database – NHB (2014)
2013-14	1077.7	11147.1	Horticulture Database, NHB, New Delhi, 2014
2016-17	985.0	11419.00	NHB, New Delhi, 2017-18
2017-18	1009.0	12510.00	NHB, New Delhi, 2018

Table 2.3: Area and Production of Major Citrus Types in India

Citrus Type	Area, '000 ha				Production, ' 000 metric tonne			
	1997-98	2001-02	2012-13	2017-18	1997-98	2001-02	2012-13	2017-18
Mandarin Orange	165.4	198.9	311.2	430	1472.4	1660.1	2906.3	4982
Sweet Orange	67.4	126.4	323.2	183	881.8	1210.4	3519.9	3369
Acid lime and lemon	129.8	161.3	255.2	236	110.1.3	1413.7	2523.5	2506
Kinnow	-	19.5	-	-	-	235.3	-	-
Others		112.3	152.9	157		269.6	1140.0	1654
Total	487.7	618.5	1042.5	1006	4310.7	4789.1	10089.7	12510

Source: Database, 2009 and 2017of the National Horticulture Board (NHB), India.

At present, citrus production is 125.10 lakh tonnes, with maximum share of mandarin (49.82 lakh tons) followed by sweet orange (33.69 lakh tons) and acid lime (25.06 lakh tons). The production trend during last 15 years clearly established considerable production gains both in sweet orange and limes and lemons.

Table 2.4: Statewise Area, Production and Productivity of Lime/Lemon

Area in '000 ha, Production in '000 metric tonne and productivity = tonne/ha

State	2016-17		
	Area	Production	Productivity
Andhra Pradesh	35.65	587.09	16.47
Karnataka	12.71	301.47	23.72
Maharashtra	23.67	203.93	8.62
Madhya Pradesh	10.47	123.22	11.77
Bihar	18.00	128.70	7.15

State	*2016-17*		
	Area	*Production*	*Productivity*
Assam	15.44	139.49	9.03
Jharkhand	4.84	47.74	9.86
Chhattishgarh	6.91	110.90	16.05
Others	119.82	841.74	7.03
Total	247.51	2484.28	10.04

Table 2.5: Statewise Area, Production and Productivity of Orange (Mandarin)

Area in '000 ha, Production in '000 metric tonne and productivity = tonne/ha

State	*2016-17*		
	Area	*Production*	*Productivity*
Punjab	49.356	1140.31	23.11
Madhya Pradesh	69.70	545.64	7.83
Maharashtra	107.70	775.12	7.20
Rajasthan	15.50	297.00	19.16
Assam	17.43	224.89	12.90
Karnataka	4.09	94.81	23.18
Nagaland	6.47	54.42	8.41
Meghalaya	9.15	45.32	4.95
Manipur	4.86	42.97	8.84
Tripura	7.68	31.40	4.09
Others	88.26	340.15	3.85
Total	380.20	3592.03	9.45

Table 2.6: Statewise Area, Production and Productivity of Mosambi (Sweet orange)

Area in '000 ha, Production in '000 metric tonne and productivity = tonne/ha

State	*2016-17*		
	Area	*Production*	*Productivity*
Andhra Pradesh	76.11	1238.80	16.28
Maharashtra	53.94	643.19	11.92
Karnataka	1.76	28.25	16.05
Punjab	2.72	22.42	8.24
Others	72.08	868.83	12.05
Total	206.61	2801.49	13.56

In India, there are 29 states involved in citrus production but 9 states cover more than 67 per cent of area accounting for 80 per cent of total production. There exists very high degree of disparity in area and production vis-a-vis productivity. For example, the Maharashtra state ranks first in area but in production it is second and in productivity it is far below. In spite of poor yield in Maharashtra, there was almost four fold increase in area under citrus. The high market price and steady increase in demand in the domestic market keeps citrus cultivation a highly remunerative venture. Large area of citrus in Maharashtra has been planted recently and yet to come to bearing and that may be one of the reasons for lower productivity reflected in case of Maharashtra.

Rapid area expansion in sweet orange in last 2-3 years period signifies regaining of the confidence building process in the minds of citrus growers. Sweet orange has once been considered as most susceptible citrus type to decline. The area expansion trend in sweet orange is clearly noticeable in Andhra Pradesh. In case of acid lime, which was once under threat due to viral (Tristeza) and bacterial (citrus canker) diseases, also recorded high growth in Andhra Pradesh, Maharashtra and Madhya Pradesh.

Commercial Citrus Growing Belts

There are four major citrus growing belts where various citrus cultivars adapted to diverse climate and soils conditions. The major citrus species grown in India are:

1. **Mandarins (*Citrus reticulata* Blanco):** Nagpur mandarin, Khasi mandarin, Coorg mandarin, Kinnow mandarin
2. **Sweet oranges** (***Citrus sinensis*** **Osbeck):** Mosambi, Sathgudi, Blood Red, Malta, Hamlin, Pineapple, Valencia and Jaffa
3. **Acid lime (*Citrus aurantifolia* Swingle):** Kagzi lime, Vikram, Pramalini, PKM-1, Baramasi Kagzi seedless lime
4. **Lemon (*Citrus limon* Burm.):** Baramasia, Assam lemon, Pant lemon, Gandhraj, Italian lemon and Eureka lemon
5. **Grapefruit (*Citrus paradisi* Mad.):** Duncan and Marsh seedless Ruby
6. **Pummelo (*Citrus grandis* Osbeck):** Chakotra, and local selections

Some special features of Indian citrus industry and notable changes took place in recent times are as follows:

i. The entire Khasi mandarin cultivation of north-eastern region is of seedling origin. The quality of fruits is excellent but due to absence of proper cultivation practices, a majority of citrus orchards are in various stages of decline in the entire region.

ii. The once famous Coorg mandarin in the southern state of Karnataka is under the threat due to severe infection of greening disease. In this region, citrus plants are used more as shade/cover crop to coffee plantation. A local strain of *Trichoderma* sp. was found to be effective against phyotphthora diseases in Coorg mandarin.

Nagpur Mandarin

iii. In the plains of north-western India, sweet orange and mandarins, constitute the bulk of citrus fruits. The mandarin cultivar, Kinnow occupies a predominant position in the states like Punjab, Haryana, and Rajasthan and has become a highly successful commercial fruit crop. Advanced cultivation practices including drip irrigation are followed. The budded plants on rough lemon (Jatti Khatti) rootstock are highly productive. Export of Kinnow fruits to some of the neighboring countries has been successfully attempted from this region.

iv. The Vidarbha region of Maharashtra has the largest concentration of Nagpur mandarin plantation in India. This is a seeded variety. Recently, ICAR-CCRI, Nagpur has identified a seedless Nagpur mandarin with no seed or 1-2 seeds and planting material of this variety is distributed to the farmers.

v. Sweet orange suffered decline in north Indian state like Punjab. Presently sweet orange Mosambi in Maharashtra and 'Sathgudi' in Andhra Pradesh

Sweet Orange

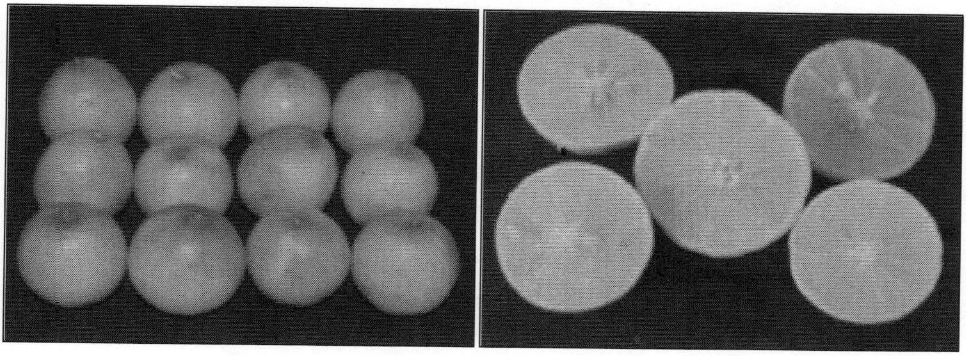

Acid Lime

are cultivated commercially. Virus cleaning through STG, budwood certification programme for production of disease free planting material and cross protection for CTV through isolation and inoculation of milder strains of virus are some of the researchable issues that should receive priority attention in sweet orange.

vi. In acid lime which is the third most important fruit crop in India, a several high yielding varieties like Parmalini, Vikram, Jai Devi, Sai Sarbati (selections) have been released. Also, a lime x lemon hybrid (ALH-77, a canker tolerant hybrid) has been developed, showing very low incidence of citrus canker. In North India lemon cv Eureka and lemon-citron-type named as Assam lemon are commercially successful. Recently, CCRI, Nagpur has identified very promising high yielding clones of acid lime *viz.* Acid lime-7 and acid lime-8. Planting material has been given to growers on large scale.

Problems of Citrus Industry

A number of problems namely, general neglect; irregular bearing; scarcity of water for irrigation; poor water management practices; cultivation in unsuitable soils and Salinity; Improper fertilizer schedule, incidence of citrus blackfly "Kolshi", psylla, leaf miner, bark eating caterpillar, trunk borer, mealy bugs, thrips and mites; heavy incidence of *Phytophthora,* a root rot and gummosis causing fungus in nurseries and grown-up orchards; presence of virus and virus like diseases; canker in acid lime, bacterial and greening diseases; improper handling without any postharvest treatment leading to reduced shelf life of fruits are prominent ones which must be addressed.

The region wise problems of various belts of citrus industry are also different.

North-eastern Region

Khasi and Darjeeling mandarins are the major cultivars in addition to lot of variability in germplasm of citrus. The major problems of citrus cultivation in the region are:

- ☆ Heavy and long spells (upto 6-7 months) of rainfall (April-October)
- ☆ Heavy soil erosion due to cultivation on 60-70° slopes
- ☆ Plantation of seedlings
- ☆ Acidic soils
- ☆ Wide spread zinc deficiency
- ☆ Malnutrition -major and minor elements
- ☆ No control against insect pests and diseases
- ☆ Prevalence of tristeza and greening
- ☆ Shortage or no irrigation during November to March

North-western Region

Kinnow predominates in this region and some extent Sweet oranges Jaffa, Malta are taken. This region comprises Fazilka, Abohar, Faridkot, Bathinda and Hoshiarpur districts of Punjab, Sirsa of Haryana and Sriganganagar Distt. of Rajasthan and Sirmour and Kangra districts of Himachal Pradesh. Recently, Nagpur mandarin area has expanded in Rajasthan mainly in Jhalawar and Kota Distt. Acid lime dominated in Khaira Distt of Gujarat and falls under arid irrigated climate Zone with increased salinity and high soil pH (8.0-9.0). Citrus is sensitive to excess moisture conditions within its rootzone and thrives best on well drained soils. Other major problems are:

- ☆ Kinnow heavy feeder of nutrient and water because of shallow rooting system and prolific bearing
- ☆ Flood irrigation acting as a predisposing factor for *Phytophthora* infestation
- ☆ Presence of hard pan
- ☆ Lime induced micronutrient deficiency of N, Ca, Fe, Mn, Cu and Zn
- ☆ Salinity and high soil pH (8.0-9.0)
- ☆ Incidence of insect pests such as psylla, mealy bugs, BECP, mites *etc.*
- ☆ Presence of tristeza and greening and *Phytophthora* caused diseases.

Central Region

This zone includes mainly Nagpur, Amravati, Wardha, Yavatmal, Buldana, Akola districts of Vidarbha in Maharashtra and adjoining Pandhurna and Saunsar Tahsils of Chindwara and Shajapur and Mandsaur districts of Madhya Pradesh. Some mandarin growing soils of the region having higher clay content mainly concentrated in sub-surfacc, low coarse fragments of sand and $CaCO_3$ upto 150 cm depth are not suitable for Nagpur mandarin cultivation from regular flowering point of view. The other major factors are:

- ☆ Irregular flowering due to high clay content in the sub-surface
- ☆ Non-maintenance of proper crop load
- ☆ Incidence of citrus blackfly 'Kolshi', psylla, Bark eating caterpillar (BECP), mites *etc.*

- ☆ Wide scale incidence of *Phytophthora* root rot and gummosis
- ☆ Presence of virus and virus like diseases and greening diseases
- ☆ Canker problem in acid lime

Southern Region

It includes parts of Andhra Pradesh, Karnataka, Tamil Nadu and Kerala. Sathgudi sweet orange, Coorg mandarin and acid lime are the main citrus cultivars of this zone. Inadequate nutrition level of orchards in form of Zn deficiency was found one of the main reasons for poor performance in most orchards of this region. The other major factors are:

- ☆ More rainfall and humid climate in hilly region
- ☆ Presence of hard pan in some parts of Andhra Pradesh with high pH and water scarcity
- ☆ Acidic soil pH including calcium deficiency in hilly areas
- ☆ Incidence of insect pests such as Scales, Psylla
- ☆ Greening, tristeza and *Phytophthora*
- ☆ Rainfed cultivation in hilly areas

Technologies Available with ICAR-Central Citrus Research Institute, Nagpur

- ☆ Shoot tip grafting technique for cleaning of planting material to make it disease free
- ☆ Complete protocol for production of disease free planting material (Bud wood certification)
- ☆ Leaf sampling technique for reliable foliar analysis and Leaf nutrient standards for optimum production
- ☆ Fertilizer requirement for pre-bearing and bearing Nagpur mandarin and acid lime
- ☆ Weed control in citrus orchards upto 300 days in a year using herbicides
- ☆ Control of fruit drop
- ☆ Bahar treatment and Crop regulation
- ☆ Chemical control of citrus blackfly, psylla, leaf miner, BECP, mites *etc.*
- ☆ Mass multiplication technique of predator *Mallada desjardinsi* (Navas) for bio-control citrus insect pests
- ☆ Chemical control of *Phytophthora* causing foot and root rot and postharvest disease
- ☆ Mechanical sorting, washing, waxing and size grading of citrus fruits
- ☆ Prolonging of harvest of Nagpur mandarin using gibberellic acid
- ☆ Maturity standards for citrus fruits
- ☆ Degreening technique for Nagpur mandarin and sweet orange

- ☆ Evaporative cool chamber for field level storage of citrus fruits
- ☆ Packing of citrus fruits using vented corrugated boxes for distant transport.

Disease Free Quality Planting Material

Disease free nursery stock is the pre-requisite for establishing a viable and productive citrus industry. Six to eight million plants of Nagpur mandarin are being multiplied every year in about 350 nurseries located in Amravati and Nagpur districts of Maharashtra and are being supplied to all over country without any certification for purity and disease-free condition. Hence, many of the diseases are spreading like fire to virgin areas due to sheer ignorance of multiplication from uncertified budwood for future multiplication. Strict quarantine and budwood certification programme must be started on the exact lines of citrus Variety Improvement Programme of USA, Spain, Brazil, *etc.* at district level to supply the healthy plant material to growers. Central Citrus Research Institute, Nagpur is producing disease free planting material of Nagpur mandarin, sweet orange 'Mosambi' and acid lime seedlings and making it available for distribution.

Soil and Nutrition Management

The fertilizer consumption per unit area is lowest in India and more so in fruit crops. The National Commission on Agriculture has reported that on an average less than 0.5 kg fertilizer per tree is applied to fruit crops. The knowledge on precise macro and micro nutrient requirements of different citrus cultivars is a pre-requisite for fertilizer use efficiency. The soil and foliar application of micronutrients are used for correcting the deficiency with varying success. Zinc sulphate proved ineffective under calcareous conditions while chelated source like Zn-EDTA@ 50 g Zn/plant was found to be very effective in alleviating Zn deficiency.

Efficient Use of Water

Fertigation technique of applying nutrients and water simultaneously holds a big promise in citrus orchards. The water requirement of Nagpur mandarin was determined as 651.0, 849.0 and 997.3 mm/year for young, middle age and mature trees, respectively, using the predicted potential evapo-transpiration based on meteorological data. The highest bio-mass per plant was obtained with 18-19 irrigation as scheduled at 0.05 Mpa SWPusing 182.4 cm water/tree/annum.

Drainage in Nagpur Mandarin Orchards

Soil texture and depth are two important factors for deciding drainage problems for Nagpur mandarin cultivation. Very deep black soils (depth > 6 ft) were found to be unsuitable for mandarin mainly due to lack of drainage and thereby improper stress to the plant for induction of flowering. Planting on raised bed system found to be beneficial as it always keeps rooting zone free from water logging, increases rooting depth in shallow soils and also acting as a soil and water conservation measure.

Citrus based Intercropping

The ultimate aim of intercrops is to obtain additional income from the same piece of land and to improve the soil fertility status with emphasis on the health of citrus trees. Unsuitable intercrop depletes soil fertility and affects the growth. It was profitable to grow cotton as an intercrop during the first three years provided a safe distance is maintained from the tree. The vegetable crops like peas, turnips, cabbage and knol khol have been recommended for rabi season. Intercropping with maize, ragi, buck wheat, ginger and turmeric had undesirable effect on mandarin trees in North-east region. In the plains pulses like soybean, green gram, black gram, cowpea, gram and ground nut were found to be the most beneficial intercrops.

Citrus Protection

Insect Pests

Epidemic of citrus blackfly (*Aleurocanthus woglumi* Ashby) on Nagpur mandarin during the seventies, till early nineties had brought havoc on the citrus industry of Central India. Widespread growth of black sooty mould (*Capnodium sp.*) on the honeydew secreted by numerous citrus blackfly nymphs covering the entire tree is locally known as "Kolshi", not only inhibits photosynthetic activity but also affècts the appearance and quality of the fruit.

Citrus psylla (*Diaphorina citri* Kuwayama) is another serious pest of young flushes that causes heavy flush drop affecting the fruit set extensively. It is also an active vector of "Greening" pathogen known to cause slow decline. Similarly, brown citrus aphid (*Taxoptera citricida* Kirkaldy) is the efficient vector of tristeza virus that causes "quick decline" of the tree. Bark eating caterpillar (*Inderbela* sp.) is thriving in ignored or old citrus orchards. Citrus trunk borer (*Monohammus versteggi* Nitzema) is a serious pest of Khasi mandarin. A single borer can kill one tree. As high as 40-60 per cent orchards suffer from this pest.

Virus and Virus like Diseases

More than 15 viral diseases of citrus have been reported from India. Among them tristeza, psorosis, crinky leaf, infèctious variegation, ring spot, mosaic and Exocortis are major ones. Planting of virus free planting material is the basic need of the industry: Virus elimination may be achieved either through shoot tip grafting or by nucellar plants. Budwood thermotherapy also eliminates the viruses. In India, CTV cross protection was initiated in early 1970s. Though initial five to six years it was giving promising results in terms of growth of acid lime plants after wards it did not perform upto satisfaction.

Bacterial Disease

The canker is major bacterial disease in citrus. Pruning of infected twigs before monsoon and spray of 1 per cent Bordeaux mixture was found effective. Streptocycline in combination with Bordeaux mixture or COC also gives effective control.

Use of Resistant Rootstocks

The primary reason for shifting from seedling to budded plants was the appearance of *Phytophthora* foot rot in the Azores islands in 1842. Initially sour orange dominated as *Phytophthora* tolerant rootstocks in citriculture but later due to viral susceptibility of this rootstock, search for new rootstocks became an important researchable aspect. In India, more than 80 per cent citrus plants are being raised on Rough lemon and rest on other rootstocks (Rangpur lime *etc.*) These rootstocks are susceptible to *Phytophthora*. No rootstock is immune to *Phytophtora,* however rootstock trials for last 50 years have given good indication for region wise use of particular rootstock. Only Limocravo Brazil, Rangpur lime Knorr, Souranthan strain of Rangpur lime and Jattikhatti strain of rough lemon have shown better ranking for tolerance. Trifoliate orange and its hybrids are highly tolerant to *Phytophthora*.

Nematodes

Citrus nematode, *Tylenchulus semipenetrans* is the most widely distributed nematode in all the citrus growing areas. But nematodes are not the major problem in most citrus growing belts of the country except few pockets.

Postharvest Technology

Post-harvest factors affecting fruit keeping quality

Harvesting

Citrus fruits are very susceptible to damage during harvesting and proper harvesting techniques minimizes bruises and mechanical injuries. Fruits should be best harvested by using a shear, clippers *etc.* The harvested fruit should be collected either into collecting baskets or lined ventilated plastic filled crates. Collecting baskets are either tied around the waist or put over the shoulder and made with a quick opening bottom. When collecting basket is filled, it should be lowered with a rope to the ground and fruits should be emptied into a larger field container carefully not tossed.

Mechanized Sorting, Washing, Wax Coating and Grading of Mandarin Fruits

Research on mechanized handling of citrus fruit started at NRCC during 1989-90 first time in the country to study the effects of mechanized handling on citrus fruit especially 'Nagpur' mandarin. Damage to fruit was evaluated and ways to minimize damage were developed. The stage of fruit maturity fit for handling on packing line, drying temperature and time for optimum drying of wax coating were worked out. In order to reduce water loss and to improve appearance, various coatings have been tried.

Process of Pre-cooling

Among various methods of pre-cooling evaluated *viz.*, forced-air cooling, hydro-cooling and room cooling, the method of forced -air cooling is found to be 5 times faster than room cooling.

Standardization of Refrigerated Storage Conditions

Storage conditions (6-7 °C temperature and 95 per cent RH) have been standardized for long term storage. Temperatures below 4°C cause chilling injury to mandarin fruit. After refrigerated storage for 45 days, fruit has shelf life of 6 days, however, if refrigerated storage is extended to 60 days fruit has shelf life of 3 days at ambient condition. After 30 days in refrigerated storage fruits can be kept outside for 9 days.

Coatings for Enhancing the Shelf Life of Citrus Fruits

Edible Coatings Characteristics

Edible coatings should have the following characteristics:

1. The coating should be water-resistant so as to remain intact and to cover all parts of a product adequately when applied.
2. It should not deplete oxygen or build up excessive carbon dioxide.
3. It should reduce water vapour permeability.
4. It should improve the appearance, maintain structural integrity, improve mechanical handling properties, carry active agents (antioxidants, *etc.*) and retain flavour of the produce.

Different Types of Edible Coatings

Edible coatings may be composed of polysaccharides, proteins, lipids or a blend of these compounds. Their presence and abundance determine the barrier properties of material with regard to water vapour, oxygen, carbon dioxide and lipid transfer in food systems.

- Polysaccharide based coatings and films
- Protein based coatings
- Lipid based coating

The major benefit of edible coatings is that they can be consumed along with food, can provide additional nutrients, may enhance sensory characteristics and may include quality enhancing antimicrobials.

Packaging

Packaging fresh fruits is one of the important steps in the long and complicated journey from grower to consumer. Bags, crates, hampers, baskets, cartons, bulk bins, and palletized containers are convenient containers for handling, transporting, and marketing fresh produce. Packing and packaging materials contribute significant cost to the produce industry; therefore it is important that packers, shippers, buyers, and consumers have a clear understanding of the wide range of packaging options available.

Proper packaging of a product can reduce not only bruising and crushing, but can also improve marketing of produce, reduce moisture loss, prevent contamination of the product with spoilage organisms, reduce pilferage, maintain a sanitary environment during marketing. The ultimate goal of the packaging must lead to easier handling of the produce, a better quality and better marketable product.

Characteristics of Packaging Material

Different types of packages are used throughout the world, many of which have been carefully evaluated with respect to produce and market system, while other types have often adopted for general use without thorough evaluation. Changes to improve such packages are still required. Some different types of package include:

1. Sacks: flexible, made of plastic or jute.
    i. Bags: small size sack
    ii. Nets: sacks made of open mesh
2. Carton or fibreboard boxes.
3. Plastic crates.
4. Pallet boxes and shipping containers.
5. Baskets: made of woven strips of leaves, bamboo, plastic *etc.*

Sacks and Nets

The materials used for sacks and nets may be woven natural fibre (jute, kenaf, sisal, cotton), woven synthetic (polypropylene, polyethylene), knitted natural fabric (cotton), knitted synthetic (polyethylene) or non-woven synthetic (propylene).

Corrugated Fiberboard

Corrugated fiberboard (often mistakenly called cardboard or pasteboard) is manufactured in many different styles and weights. Because of its relatively low cost and versatility, it is the dominant produces container material and will probably remain so in the near future. The strength and serviceability of corrugated fiberboard have been improving in recent years.

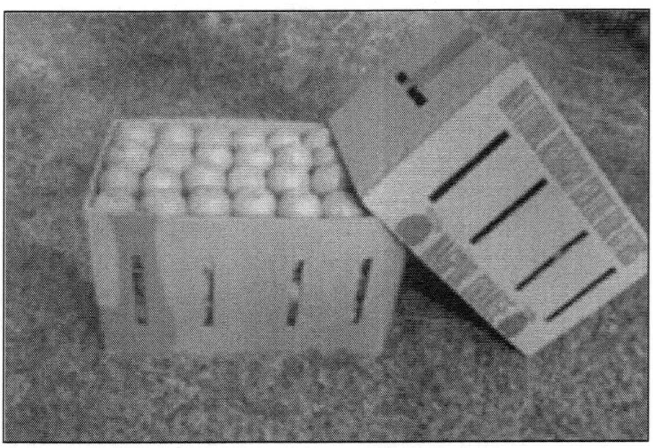

Plastic Bags

Plastic bags (polyethylene film) are the predominant material for fruit consumer packaging. Besides the very low material costs, automated bagging machines further reduce packing cost. Film bags are clear, allowing for easy inspection of the contents, and readily accept high quality graphics. Plastic films are available in a wide range of thicknesses and grades and may be engineered to control the environmental gases inside the bag. The film material "breathes" at a rate necessary to maintain the correct mix of oxygen, carbon dioxide, and water vapor inside the bag.

Shrink Wrap

One of the newest trends in produce packaging is the shrink wrapping of individual produce items. Shrink-wrapping has been used successfully to package a variety of tropical fruit. Shrink wrapping with an engineered plastic wrap can reduce shrinkage, protect the produce from disease, reduce mechanical damage and provide a good surface for stick-on labels.

Nanotechnology in Preservation and Packaging of Fruits/Fruit Products

In relation with extension of horticultural products shelf life, nanotechnology can help us in many ways, for *e.g.* controlling growth and development of micro organisms, introducing a new generation of packaging coverage's (Films) and controlling influence of gases and the harmful rays (UV), increasing strength, quality and packaging using the multiple chips (Nanobiosensors) for labeling products that considered as fundamental step to automated control of storages.

Nano packaging provides an attractive alternative to improve preservation quality of the fruits during extended storage. Blending polyethylene with nano-powder like nano- Ag, Kaolin, anatase TiO_2 has been reported to expand shelf life and improve preservation quality of fruit and its value added products.

Postharvest Management Issues in India

Postharvest management is still in infant stage in India and there is a tremendous scope to develop infrastructure in this sector. Fresh food consumption has been the traditional habit due to ample availability of food in different seasons due to tropical climate. In fresh produce, losses are very high depending on commodity, seasonal harvest, method of harvest and handling and pre-harvest practices. Fresh produce is handled in conventional way disregarding bruising injuries and infection. There are very meager facilities of cold-chain. Marketing system is not organized at many places. Farmers, traders' transporters and retailers are not aware about postharvest handling needs due to lack of education in this area.

Perspective of Citrus Industry

The demand of citrus fruits is likely to be increased by nearly 1.5 fold in 2030 and 2 fold in 2050 with an estimated population of 135 crores by 2030 AD and 160 crores by 2050. Demand is also increasing for fruits due to improving economy, better health awareness and changing lifestyle among the people. With increased productivity of 12.5-15 tonnes/ha, per capita availability of citrus fruit is also expected to be higher at 8.80-9.00 kg/year as compared to 7.10 kg/year at present.

Table 2.7: Projected Area and Production of Citrus Fruits

Citrus Cultivars	*Area (lakh ha)*				*Production (lakh tonnes)*			
	1992-93	*2012-13*	*2030*	*2050*	*1992-93*	*2012-13*	*2030*	*2050*
Mandarin	1.66	3.15	3.20	4.80	13.40	29.06	35.14	72.00
Sweet orange	1.10	3.23	3.50	4.00	8.93	35.19	64.20	60.00
Acid lime and Lemon	0.93	2.55	3.40	3.80	7.46	25.23	46.57	57.00
Other citrus	-	1.52	-	1.70	-	-	-	16.90
Total	3.69	10.42	10.1	14.30	29.79	100.89	145.91	205.90

Source: Area and Production, Horticultural Information Service NHB, November 1994 and 2010.

To achieve these goals it is estimated that the minimum annual growth of 5 per cent will have to be achieved in next 15-35 years so as to meet the domestic demand and enter in exports to tap foreign countries. With annual growth rate of 10 per cent in past *i.e.* during 1992-93 to 2013-14, it appears that estimated growth rate of 5 per cent can be easily achieved in future. With gradual increase of productivity per hectare from present level of 10 ton/ha to 15 ton/ha during next 15-35 years, the target of production can be achieved.

Chapter 3

Prospects and Challenges in Seed Spices

Gopal Lal

Member
Cauvery Water Management Authority,
Ministry of Jal Shakti (GoI), New Delhi

Introduction

India is 'The Land of Spices' and the glories of Indian spices are known throughout the world. Arid and semiarid parts of the India (Rajasthan and Gujarat) are known as **Seed Spices Bowl** and contribute ~80 per cent of total seed spices production. Total 109 spices are listed by ISO and 63 spices are grown in India and out of which 20 are being classified as seed spices. Out of 20 seed spices, ICAR-NRCSS working of 10 most important seed spices namely cumin, coriander, fennel, fenugreek, ajwain, anise, caraway, celery, dill and nigella (NRCSS Vision, 2050). Seed spices also grown in some other states on considerable areas such as Uttar Pradesh, Madhya Pradesh, Bihar, West Bengal, Odisha, Punjab, Karnataka and Tamil Nadu. Since there is a large scope of seed spices by introducing them in new areas, the higher yields can also be achieved effortlessly by implementing new technologies. Among seed spices, maximum area is under cumin followed by coriander, fennel fenugreek, ajwain, celery, nigella, dill, anise *etc.* Being seasonal crops, these are grown extensively in rotation with food crops and also as inter/ mixed crops under rainfed/irrigated conditions. All these crops mostly grown in rabi season. India is the largest producer, consumer and exporter of seed spices in the world. Seed spices contribute ~50 per cent of total area and ~19 per cent of production of total spices in the country. The area under seed spices is about 1.79 million hectare and production is about 1.86 million tons (DASD, Calicut- 2017-18). Fist time in the seed spice history the productivity has crossed 10 quintal ha. India is consistent source of seed spices for importing countries worldwide. There has been ever increasing demand of seed spices and importing countries look at India for quality produce of seed spices.

Seed spices play a significant role in our national economy because of its large domestic consumption and growing demand for export. India is exporting ~15 per cent of its production annually and full fill nearly 50 per cent of world demand. The total export of seed spice crops is Rs 3738 crore, out of which cumin alone contributes Rs 2418 crore annually (Spice Board India, 2017-18). Other than India, seed spices are grown in different part of the world covering mainly Mediterranean region, South Europe and Asia. Morocco, Russia and Bulgaria are the major producing countries for coriander, Turkey, Iran and Egypt for Cumin; Egypt, China, Romania and Russia for fennel; Morocco and Bulgaria for fenugreek; Iran and Egypt for Ajwan; Germany and Hungry for dill; southern France and china for celery; Bulgaria, Cyprus, Germany, and Russia for anise; Pakistan, Sri Lanka and Egypt for Nigella.

Increase in Area Production and Productivity

The significant growth in area, production and productivity of major seed spices were experienced in past 15 years (Figure 3.1). The area under seed spices was 9.02 hectare during 2002-03, which has increased to 17.9 lakh hectare during 2017-18; likewise the production was 4.11 lakh tons during 2002-03 which has jumped to 18.6 lakh tons during 2017-18. Productivity increased more than two times *i.e.*, from 456 kg/ha (2002-03) to 1039 kg/ha (2017-18) (Figure 3.1).

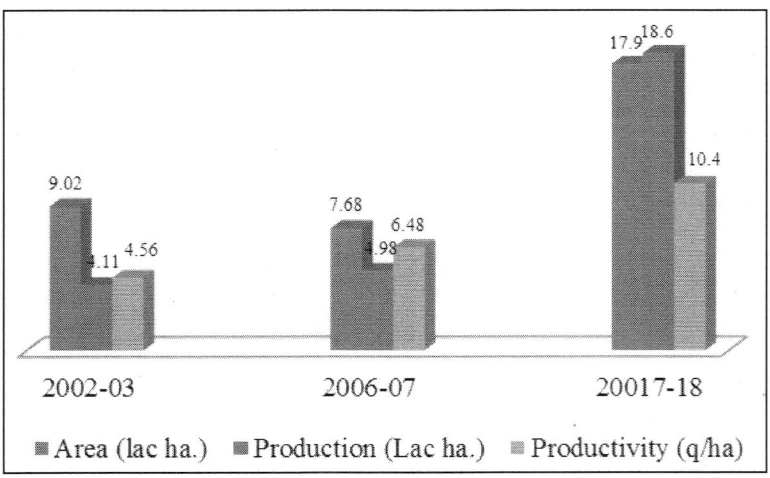

Figure 3.1: Area, Production and Productivity of Seed Spices (*Source*: DASD, Calicut- 2017-18).

Present Status of Seed Spice Research in India

Systematic research wok started when the Government establishment the All India Coordinated Spices and Cashew Improvement Project (AICSCIP) in 1971. The biggest milestone was achieved in 1975 when cumin, coriander, fennel and fenugreek were added in the AICSCI project by starting four new seed spices research centres, *viz.* at Coimbatore (Tamil Nadu),Guntur (Andhra Pradesh), Jagudan (Gujarat) and Jobner (Rajasthan). In 1986, two organizations came into existence, first, All India Coordinated Research Programme on Spices at Calicut from AICSCIP for research and second, Spice Board of India by merging Cardamom Board (1968) and Spice Export Promotion Council (1966) for promotion of export and implementation of various government schemes on spices including seed spices (Singh and Solanki, 2015). The landmark achievements were made after the exclusive set up of institutes for R and D of spices and seed spices. ICAR-Indian Institute of Spices Research Calicut (estb. 1986) and National Research Centre on Seed Spices Ajmer (estb. 2000) have pioneered in developing a number of high yielding varieties and production technologies. AICRP Spices is also conducting location and situation specific research on spices including seed spices at 14 centers in 12 states. During ninth plan period, ICAR established National Research Center on Seed Spices at Ajmer in April, 2000 to conduct basic and strategic research exclusively on seed spices crops to further generate technology for enhancing the production and productivity.

Research Achievements in Seed Spices

1. Crop Improvement

In the National Agriculture Research System for seed spices includes NRC on Seed Spices and AICRP Spices, the germplasm of seed spices have been evaluated for yield, adaptability and the reaction to diseases and pests, insects-pests, maturity, adaptability, processing attributes and nutritional qualities, maintained scientifically and updated with new accessions. The AICRPS has contributed substantially by enriching germplasm, developing high yielding varieties; technologies for increasing the production and productivity; combating pests and pathogens to reduce crop losses, secondary agriculture *etc.* The total germplasm collection available at NAGS is more than 2200 in 1o seed spice crops. Efforts to update the existing germplasm through the addition of indigenous and exotic collection need to be intensified (Verma *et al.,* 2018). There is also a strong need to research for new genes in land races and wild type to develop stress tolerant varieties. So far, about 100 varieties have been developed in seed spices by different institute and universalities (www.seednet.gov.in). ICAR-NRCSS, Ajmer alone developed 23 varieties in different seed spices crops.

2. Quality Seed Production of Seed Spices

Seed replacement rate in seed spices is very low (15 per cent) which is a major challenge for enhancing production and productivity of seed spices. Providing good quality seeds is one of the most important and easiest means to accelerate the productivity of seed spices in the country. Importance of quality seed includes seed is a vital input in crop production, it is the cheapest input in crop production and key to agriculture progress, crop status largely depends on the seed materials used for sowing, response of other inputs in crop production depends on seed material used, the seed required for raising crop is quite small and its cost is so less compared to other inputs, this emphasis the need for increasing the areas under quality seed production and it is estimated that good quality seeds to improved varieties can contribute about 20-25 per cent increase in yield (Lal, 2018).

3. Production Technology of Seed Spices

The agro-techniques are highly site specific and vary for different crops. Suitable package of practices including, efficient water, optimum sowing time, seed rate, nursery management, transplanting, spacing, intercultural practices, weed management, cropping system, intercropping and organic farming practices has been standardized for different seed spices crops for realizing the optimum potential of improved varieties in the different agro-climatic conditions. The production technology for off season cultivation has also been worked out for leafy coriander and fenugreek production. In coriander, plastic covered walk-in-tunnel and shade net covered walk-in-tunnel with low pressure drip system was found highly suitable for off-season cultivation for green leaves (Lal, 2018). Raised bed technology with drip fertigation and plastic mulching has been found effective for fennel, dill, celery and ajwain (Annual Report, 2012-13, 13-14). Irrigation efficiency is only to the tune of 30-35 per cent, therefore only means to provide the water is effective management and enhancing the water use efficiency to the level of 80 per cent which is possible with the use of micro irrigation. Among various methods tried drip irrigation has proved successful in exhibiting high water productivity by saving irrigation water from 25-60 per cent along with 10-60 per cent increase in yield as compared to conventional method of irrigation. Different cropping system has been helpful for crop diversification, efficient utilization of resources and making intensive use of inputs for sustainable income to marginal farmers. The diverse favourable agro-climatic conditions of the country provides opportunity to cultivate variety of horticultural crops such as fruits, vegetables including root and tuber crops, flowers, plantation and spices crops and medicinal and aromatic crops. These crops provide additional income to farmers. Intercropping of fennel or with knolkhol in 1:2 ratio followed by fennel/dill with cabbage in 1:1 intercropping ratio is very effective for realizing higher yield and benefit. Cluster bean- cumin-summer moong followed by pearl millet- cumin- summer moong cropping sequence is very effective cropping sequence for cumin growing farmers. Intercropping of seed spices in arid fruit orchards has also been assessed and found economical (Mehta *et al.*, 2017). The use of bio-fertilizers, VAM fungi, biological nitrogen fixers and other beneficial microbial agents also need to be optimally used to attain an efficient management in seed spices. Seed spices can be accommodated in other crops in different cropping and/or farming system. The fenugreek can be grown as a fodder crop which is rich in protein. The straw from different spices can be used as a feed for the animal by blending it in different ratio. Seasonal vegetables and flowers can be rotated intercrop with seed spices (Lal and Verma, 2018). In recent years, organic farming has gained considerable importance. The spices produced from this type of farming are without any harmful ingredients.

4. Protection Technology

Among various insect pests associated with seed spices crops, sucking pests comprises the largest group and causes maximum damage to the crop at field level. Among sucking pests aphids are major yield limiting biotic factor, as it causes more damage than any other pest on most of seed spices crops. More than one species of aphids found attacking on the many of seed spice crops. Other sucking pests associated with seed spice crops are thrips, jassids, hopper, bugs *etc.* Seed wasp is another important pest of both field and storage and is important quarantine pests. Defoliators are also found on these crops but have limited population and damage. Most of these pests are polyphagous, attacking variety of vegetation and a few species are exclusive for seed spices. Biological control, botanical and chemical methods have been developed to control the pests in seed spices.

The disease infestation in seed spices lower the production and productivity. Hence management of disease one of the important practices which can enhance the productivity of seed spices. Diseases like wilt, blight and powdery mildew in cumin; wilt, powdery mildew and stem gall in coriander; blight and gummosis in fennel, and powdery mildew, downy mildew and mycoplasma-like organisms (MLOs) in fenugreek frequently attack these crops causing heavy loss of yield and deteriorate the quality of the produce. Though the package of practices for the control/management of each of these diseases have been formulated but still diseases like wilt (cumin and coriander), root rot (fenugreek) and gummosis (fennel) are difficult to be controlled completely. Concerted efforts are needed to evolve appropriate technology for their effective control or varieties which can tolerate or resist this disease (Lal, 2018; Khare, 2014a and b).

5. Postharvest Management and Processing of Seed Spices

The postharvest handling activities start after the harvesting of crop, first with drying of freshly harvested crop, threshing, seed drying, cleaning, grading, packing and transportation of whole seed to market, processing plant or storage place. On farm processing is an important aspect to prevent deterioration of the produce and maintain the quality. Improvements in hygiene, packing and storage facilities help in keeping quality of spice flavours and reduce chances for afla-toxin and salmonella contamination of seed spices. Many products have been developed by NRCSS in seed spices such as essential oil, oleoresins, cryogrind powder of seed spices, coriander, fennel, dill squashes, RTS, biscuits *etc.* (Lal, 2018).

6. Technologies Dissemination

Only 40 per cent farmer households in India have access to various sources of information for modern technology for farming (Adhiguru *et al.*, 2009). The popular information sources among farmers have been reported to be fellow progressive farmers and input dealers, followed by mass media. The public extension system and extension workers have been found to be accessed by only 5.7 per cent households. Only 4.8 per cent of the small farmers have access to public extension workers as compared to 12.4 per cent of large farmers. Public-sponsored other extension methods, like training programs, study tours and so on, were found to have least influence on the adoption of modern farm technologies. The ratio of extension staff to farmers varies widely across the country from 1:300 in Kerala to 1:2000 in Rajasthan (Raabe, 2008). Looking the scarcity of transfer of technology NRCSS used to disseminate the developed technology through collaboration with different KVKs, farmers fair, training programme, fields days, Front Line Demonstration (FLDs) *etc.*

Major Challenges in Seed Spices

1. The restricted gene pool, small flower size that hampers hand emasculation and pollination in seed spices. Extensive programme on collection and characterization of germplasm lines to enhance the variability and ultimately development of superior varieties in seed spices. Recombination breeding should be promoted to enhance the yield and quality in seed spices. There is strong need for development of molecular markers database and mapping population in seed spices for tagging of candidates genes.
2. Seed spice diseases such as wilt, blight and yellowing disease in cumin, stem gall problem in coriander, dwarf varieties with tolerance to *Ramularia* blight and gummosis in fennel
3. Seed replacement rate (SRR) of seed spices is very needed to be increased
4. Standardization of minimum seed standards in seed spices
5. Priming and pelleting of the seeds of seed spice to hasten the germination
6. Focused work is needed to develop GAP standards for organic and non-organic seed spice cultivation both for domestic and international need
7. PoP's for better resource management in seed spices
8. Development of climate resilient varieties
9. Pollination management in seed spices
10. Promoting production of organic seed spices
11. Mechanization, precision farming and pesticide residue management
12. Forecasting models for IDM/IPM
13. Postharvest handling and processing for value addition *etc.*
14. The challenge faced by India in the global market is lack of uniformity in the quality aspects in terms of quality standards among importing countries. However, each country that imports spices has its own stringent food safety regulations to maintain quality standards of imported products. The exporters have to follow the quality standard specified by the countries to which they are exporting so as to avoid rejection of consignments.

REFERENCES

Adhiguru P, Birthal PS and Kumara GB (2009). Strengthening pluralistic agricultural information delivery systems in India. *Agricultural Economics Research Review*, Vol. 22, pp. 71–79.

Khare MN, Tiwari SP and Sharma YK (2014a). Disease problems in the cultivation of I. Cumin (*Cuminum cyminum* L.) II. Caraway (*Carum carvi* L.) and their management leading to the production of high quality pathogen free seed. *International J. Seed Spices* 4 (1): 1-8.

Khare MN, Tiwari SP and Sharma YK (2014b). Disease problems in fennel (*Foeniculum vulgare* Mill) and fenugreek (*Trigonella foenum graceum* L.) cultivation and their management for production of quality pathogen free seeds. *International J. Seed Spices* 4(2): 11-17.

Lal G. (2018). Scenario-Importance and Prospects of Seed Spices: A Review. Curr Inves Agri Curr Res 4(2): DOI: 10.32474/CIACR.2018.04.000181.

Lal G and Verma AK (2018). Seed species for diversified agriculture to enhance farmers income. In: *Shod Chintan*- Intensification and Diversification in Agriculture for Livelihood and Rural Development (Editors: H.P. Singh, R.C. Srivastava, Dhiraj Sharma and Babita Singh), ASM Foundation, New Delhi, 10: 214-219.

Mehta RS, Meena SS and Lal G (2017). Effect of intercropping seed spices with vegetable for enhancing system profitability *International J. Seed Spices,* 7(2): 33-39.

NRCSS Annual Report, 2012-13 and 2013-14

NRCSS Vision (2050). NRCSS, Ajmer, India

Singh B and Solanki RK (2015). Status of seed spices research and development in India. *Indian Journal of Agricultural Sciences,* 85(2): 151–6.

Verma AK, Choudhary S, Meena RS, Lal G and Sharma M (2018). Improvement of Seed Spices. In: *Phytochemistry of Fruits and Vegetables* (Peter, K.V. Editor). Brillion Publishing, New Delhi– 110005, India. pp. 287-314.

Chapter 4

Grape Cultivation for Higher Income and Nutritional Security

M.K. Verma[1] *and Tanushree Sahoo*[2]

[1]*Principal Scientist and* [2]*Ph.D. Scholar*
Division of Fruits and Horticultural Technology,
ICAR- Indian Agricultural Research Institute, New Delhi

Grape is a versatile crop being the reservoir of diverse array of nutraceuticals. Grape products can serve as a potential weapon to fight against many deadly chronic diseases owing to its tremendous antioxidant activity. Good health is one of the greatest blessings of life. It is evident that, regular dietary intake of fruits and vegetables is associated with reduced risk of cardio-vascular diseases (Bazzano *et al.*, 2003). Grape is one of the ancient fruit crops known to human civilization, which holds a unique position among the fruits owing to its diverse array of nutraceuticals and multifold uses. India ranks ninth in grape production, with an annual production of 2.43 million tons from an acreage of 0.12 million ha with very high (21.10 t/ha) productivity (NHB, 2015). Because of special arbour training systems provided for grape cultivation in India, productivity is highest among the grape growing countries of the world.

Maharashtra is the leading state in production of grapes in the whole country. With regard to agricultural land under grape cultivation and grapes production, Nasik and Sangli districts are at forefront in the state. Apart from these, grapes cultivation has also been spread in the districts of Ahmednagar, Pune, Satara, Solapur and Osmanabad. Now-a-days, grapes are produced in Latur district of Maharashtra also. However, Nasik and Sangli districts are the major producer of grapes in a scientific manner. Area under grapes in Maharashtra is 86 thousand ha and production is around 774 thousand tons of grapes annually. Total export of grapes from India is 108.58 thousand tons during 2011-12 valuing of Rs. 602.88 crores, out of which, nearly 80 per cent is exported from Maharashtra.

Nutritional Importance

- Being rich source of phenolic compounds, grapes possesss powerful antioxidant properties.
- The ample consumption of grape and its by-products help to get rid of injuries caused due to evolution of reactive oxygen species.
- The juice and wine help to slow down the aging process and also boosts the metabolism of human body.
- Anthocyanins are found concentrated in the peel of berries and in some cases even in the pulp (known as teinturier varieties; example is Pusa Navrang). These compounds are having different kinds of fractions namely, Malvidin-3-glucoside, Delphinidin-3-glucoside, Petunidin-3-glucoside, Cyanidin-3-glucoside *etc.*
- Out of different fractions, 'Malvidin-3-glucoside' is found to be dominant in coloured grape cultivars.
- Anthocyanins are having anti-cancerous, anti-inflammatory and anti-diabetic properties.
- Sufficient intake of grape berries enhances memory and promotes brain health.
- The flavonoids are also found in rich quantity in grapes, which have important function in immune system of human body.
- Gallic acid, catechins, epicatechins, tannins are the different kinds of phenolic compounds in grapes.
- Grape seed is a rich source of compound called 'Resveratrol', which play a vital role in strengthening the cardio-vascular system.
- To keep the heart and blood vessels in healthy condition, grape seed powder is highly recommended, especially to the diabetic patients.
- Grape has been rightly called as 'super food for the future' due to its numerous health benefits.
- Being a potent antioxidant rich food, its addition to the food basket will safeguard against various deadly chronic diseases.

Grape Cultivation Zones in Indian Subcontinent

Grape is grown under a variety of soil and climatic conditions. In India, broadly three distinct agro-climatic zones found for grapes growing regions; namely, sub-tropical, hot tropical and mild tropical climatic regions.

Sub-tropical Region

This region covers the north-western plains corresponding to 28° and 32° N latitude including Delhi; Meerut district of Uttar Pradesh; Hissar and Jind districts of Haryana; and Bhatinda, Ferozpur, Gurdaspur and Ludhiana districts of Punjab. Vines go through a dormancy phase during winter season and bud break starts in the first week of March. In this region the major drawback is occurance of pre-monsoon shower in the first week of June, and therefore, only 90-95 days are available from the initiation of growth to harvest. Consequently, 'Perlette' is the only early ripening variety dominant in this region. Rain induced berry cracking is a problem with Thompson Seedless in this region. Single pruning and a single harvest is the general practice here.

Hot Tropical Region

This region covers Nashik, Sangli, Solapur, Pune, Satara, Latur and Osmanabad districts of Maharashtra; Hyderabad, Ranga Reddy, Mahbubnagar, Anantapur and Medak districts of Andhra Pradesh; and Bijapur, Bagalkot, Belgaum, Gulberga districts of northern Karnataka lying between 15° and 20° N latitude. This is the major viticulture region accounting for 70 per cent of the area under grapes in the country. Vines continue their growth round the year without any dormancy. Double pruning and a single harvest is the general practice in this region. Maximum and minimum temperature is 42°C and 8°C, respectively. The major problems in this region are soil and water salinity and drought. Thompson Seedless and its clones (Tas-A-Ganesh, Sonaka), Anab-e-Shahi, Sharad Seedless and Flame Seedless are the varieties grown in this region.

Mild Tropical Region

The region between 10° and 15° N latitude including Bangalore and Kolar districts of Karnataka; Chittoor district of Andhra Pradesh and Coimbatore; and Madurai and Theni districts of Tamil Nadu fall in this region. Maximum temperatures in a year seldom exceed 36°C, while the minimum is about 12°C. Principal varieties are Bangalore Blue (Syn. Isabella), Anab-e-Shahi, Gulabi (Syn. Muscat Hamburg), and Bhokri.

Vitis Vinifera varieties are prone to mildew attack and suffer losses due to unprecedented rains during flowering and fruit set in both hot and mild tropical regions. In India, there is a preference towards fresh consumption *i.e.* approximately 85 per cent of the total production. About 120,000 tonnes of Thompson Seedless and its mutants, namely, Tas-A-Ganesh, Sonaka and Manik Chaman are dried for raisins. Some 20,000 tonnes of Bangalore Blue are crushed to make juice, and 10,000 tonnes of Bangalore Blue, Cabernet Sauvignon, Chenin Blanc, Chardonnay, Merlot, Pinot Noir and Uni Blanc are crushed to process into wine.

Propagation

Vines are raised on their own roots in India. As phylloxera or nematodes is not a serious problem in India, rootstocks are not employed generally, but in recent years, the 'Dogridge' rootstock is being employed to combat soil and water salinity problems.

Multiplication on Own Roots

Hardwood cuttings are most commonly used planting material in grapes' propagation. No Government agency is involved in the multiplication and supply of rooted cuttings. Growers themselves obtain the hardwood cuttings from elite vineyards and raise their own nurseries. Well matured canes obtained in September/

October are selected. Cuttings of 4 nodes each with a thickness of 8 to 10 mm are made from the selected canes. The fresh cuttings are soaked in running water for 24 hours to leach out the water-soluble rooting inhibitors. The basal parts of cuttings are then dipped in a 2,000 ppm strong IBA solution for five minutes before planting. It is also a practice to plant the cuttings in situ when three to four cuttings prepared and treated as above are planted at each spot in the main field. Soil drenching with chlorophyriphos 0.1 per cent is a practice to safeguard the cuttings against termite damage.

Raising on Rootstocks

Hardwood cuttings of the 'Dogridge' rootstock are rooted in polybags. Rooted cuttings of this rootstock are planted in the main field during February-March. The desired scion variety is then grafted/budded on the rootstocks in the field by wedge grafting/chip budding. Wedge grafting is more common and the best time for the operation is September-October, while June-July is the suitable time for chip budding.

The best season for planting the rooted cuttings of cultivated varieties in the main field is September-October whereas for rootstocks it is February-March. Spacing generally varies with the varieties and soil fertility. For vigorous varieties it is 6 m x 3 m or 4 m x 3 m and 3 m x 3 m or 3 m x 2 m for less vigorous varieties.

Training of Vines

The most popular training systems in India are, Bower, Telephone and Flat Roof Gable systems.

Bower System

Owing to the high productive potential, bower was a very popular system of training in the past. It is highly suited for vigorous varieties like Anab-e-Shahi, Bangalore Blue and Gulabi. But in varieties like Thompson Seedless and Tas-A-Ganesh where vine vigour and excessive foliage density affects the productivity adversely, this system is not popular.

Telephone System

T-trellis is used in this system of training. With three top wires and 'T' shaped supports, the trellis looks like a telephone pole and wires and hence the name. This system is followed for moderately vigorous varieties like Thompson Seedless and other seedless cultivars in about 25-30 per cent of the vineyard area in Maharashtra. Yields in this system are less than the bower. In very hot and dry places, sunburn of the berries and of the arms are experienced in summer.

Flat Roof Gable System

Combining the advantage of bower and the extended Y systems and eliminating their disadvantages, an inter-connected Y trellis forming a flat roof gable is being adopted. This system is particularly followed for vigorous vines (vines grafted on rootstocks). The bunches are protected from direct sunlight and well exposed to sprays of pesticides. The clusters hang within the reach of the worker of an average height. Owing to these advantages, this system is gaining popularity among the growers in Maharashtra, Andhra Pradesh and Karnataka.

Pruning of Vines

Three distinct pruning practices are in vogue in relation to cropping in the three grape growing regions of the country. In the sub-tropical region, vines are pruned only once in December and the crop is harvested once. Half of the canes are pruned to renewal spurs and the rest to fruiting canes (3-4 nodes for Perlette).

In hot tropical regions, vines are pruned twice but only one crop is harvested. All canes in a vine are pruned back to single node spurs in March-May to develop canes and the canes are forward pruned in October-November for fruiting. The number of nodes retained on a cane varies with the variety and cane thickness. There is no scope to prune earlier than October and later than November due to unfavourable weather conditions.

In the mild tropical region, vines are pruned twice and the crop is harvested twice. In varieties like Gulabi and Bangalore Blue, which are fairly resistant to rain damage and in which fruit bud differentiation is not impaired by cloudy weather and rains, pruning is done at any time of the year. As a result, five crops are harvested every two years.

Since grapes are grown in areas where the evapotranspiration exceeds the precipitation, irrigation is essential. Less than 10 per cent of the vineyard areas are surface irrigated, while the rest is irrigated by drip systems. Water requirement is calculated based on the pan evaporation using 0.8 as the crop factor. Water is applied at different rates at different stages of vine growth and berry development.

Pests and their Management

The important pests of grapes in India are, flea beetles, thrips, mealy bugs and leaf hoppers.

Flea Beetles

Very severe as they scrape the sprouting buds and eat them up completely after each pruning. Damaged buds fail to sprout. Insecticides like carbaryl at 0.15 per cent, quinolphos at 0.05 per cent, dichlorvas at 0.1 per cent or phosalone at 0.05 per cent are sprayed from the fourth day until the emergence of leaves.

Thrips

Thrips attack the ovaries of flowers and newly set berries and suck sap from them. The affected berries develop a corky layer and become brown on maturity. Thrips are effectively controlled by spraying phosphamidon at 0.05 per cent, carbaryl at 0.125 per cent, phosalone at 0.05 per cent or malathion at 0.05 per cent. Prophylactic sprays of insecticides against thrips are given once in five days from the initiation of bloom to berry set.

Mealy Bugs

Suck sap from the tender shoots resulting in crinkling and stunting of the new shoots. They excrete honey on leaves and berries and sooty mold develops on the honey. Mealy bug infected bunches are unfit for marketing. Yield losses can be up to 50 per cent due to mealy bug damage. Mealy bugs are hard-to-kill insects and the package of practices for their control in India is as follows:

1. Avoid spraying broad-spectrum insecticides particularly synthetic pyrethroids.
2. Spray only dichlorvas at 0.1 per cent mixed with neem oil 0.2 per cent or tridemorph at 0.1 per cent.
3. Release cryptolaemus montrozieri beetles at 8,000-10,000 per hectare when the berries start softening. It is better to release a mixed population of grubs and adults rather than only adults.

Leaf Hoppers

The adults and young nymphs of hoppers suck sap exclusively from the lower side of the leaves. Carbaryl at 0.15 per cent, fenitrothion at 0.04 per cent, phosalone at 0.05 per cent or quinalphos at 0.05 per cent are sprayed to control this pest.

Diseases and their Management

The important grape diseases are anthracnose, downy mildew, powdery mildew and bacterial leaf spot. In recent years, *Alternaria* is also becoming a serious pathogen.

Anthracnose is prevalent in all grape growing regions of the country. The disease is characterized by small light brown or greyish black lesions on tender shoots, young leaves, flowers and young berries. Bordeaux mixture at 0.8 per cent, copper oxychloride at 0.25 per cent or carbendazim at 0.1 per cent are used to control this disease.

Downy mildew is the most devastating disease of grapes in the tropical region of the country. The disease mainly appears on the leaves, but also attacks the flower clusters and young fruits. The losses are very high when it attacks the clusters before fruit set. Entire clusters decay, dry and drop down. Properly neutralized Bordeaux mixture at 1 per cent, copper oxychloride at 0.2 per cent, Mancozeb at 0.2 per cent, metalaxyl (Ridomil Mz at 0.2 per cent) or Phosethyl-Al (aliettle at 0.2 per cent) are used against this disease.

Powdery mildew is prevalent in all the grape growing regions. It is next in importance to downy mildew in its devastating severity. The disease is characterized by the presence of white powdery (ash like) coating in patches on both sides of the leaves, young shoots and immature berries. Powdery mildew is controlled easily by wettable Sulphur formulations. A wide range of fungicides, namely, Calaxin at 0.07 per cent, Karathane EC at 0.04 per cent, Myclobutanil (Systhane at 0.05 per cent), Triademifon (Bayleton at 0.1 per cent) and Penconazol (Topas at 0.025 per cent) are used to control this disease.

Bacteria infects leaves, shoots and berries. The symptoms appear as minute water soaked spots on the lower surface of the leaves, especially along the main and lateral veins. Mostly these spots coalesce and form larger patches. Severely infected leaves give a blighted appearance. Streptocyclin at 500 ppm is used as a prophylactic spray, while Bordeaux mixture at 0.8 per cent or copper oxychloride at 0.15 per cent is used to check its spread.

Physiological Disorders

Physiological disorders associated with high temperature and low atmospheric humidity in the hot tropical region are dead arm and trunk splitting. Salinity injury is common in Maharashtra and north Karnataka. Other physiological disorders are cane immaturity, water berries, cluster tip wilting, shot berries, uneven ripening and postharvest berry drop. The eco-physiological disorders are coulure, blossom-end rot, pink berry syndrome, berry cracking and rotting.

Export Potential of Grapes

India has emerged as the major grape exporter in the global market. The country has exported 214440.6 MT of Grapes to the world for the worth of Rs.2146.22 crores/332.75 USD Millions during the year 2017-18. The major export destinations for grapes are Netherland, Russia, U.K, Germany and U. Arab Emts. Among various fresh fruits export, grape is the highest exported commodity accounting for 11.6 per cent share out of total.

To maintain various quality parameters for grapes in order to satisfy for export standards, several practices are followed.

Practices for Quality Improvement

Shoot and Cluster Thinning

Only one or two clusters are retained per cane depending upon the density of the latter. Irrespective of the number of clusters, only the apical two or three shoots are retained. In vines trained to the flat roof gable, individual shoot length is encouraged rather than the total canopy size for preventing sunburn of the berries.

Production of Loose Clusters

Pre-bloom GA sprays of 10 ppm and 15 ppm are given respectively on the 11th to 14th day after bud break for cluster elongation. Rachides of the clusters are trimmed to retain 8-10, depending on the number of leaves available per cluster. Clusters are dipped in GA solution of 30-40 ppm when 10-20 per cent of the flowers open in each cluster for berry thinning.

Increasing Berry Size

Manual means are used to supplement chemical thinning to ensure adequate berry thinning and improve the quality of grapes. Approximately 90-120 berries are retained per cluster depending upon the number of leaves available to nourish it at 8-10 berries per every leaf depending on its size. Clusters are dipped in GA solution of 40-50 ppm concentration once at 3-4 mm size of the berries and again at 7-8 mm size. When berry diameter is to be increased to more than 16 mm, clusters are dipped in a mixture of 10 ppm BA + 25 ppm GA or 2 ppm CPPU + 25 ppm GA or 1 ppm brassinosteroid + 25 ppm GA instead of GA alone at these two stages.

In addition to the treatment with growth regulators, berry size and crispiness are increased by girdling. The width and depth of girdling are 1-1.5 mm. Girdling is done at 4-5 mm diameter of the berries.

Increasing the TSS Content

Berry thinning and cluster thinning to maintain adequate leaf/fruit ratio, while girdling will ensure a TSS content of 20°B.

Harvesting

The time of harvesting and handling of bunches are the key determinant factors behind the shelf life of berries during transit and storage. To remove mis-shapen and damaged berries, sorting is performed prior to dispatch. Delayed harvest in the day builds up field heat and poses problems in precooling. The bunches need to be packed as per the specifications directed by the concerned importing agency. Ample care should be taken to keep the bloom intact to the berries, which naturally acts to enhance storability of grapes. The grape guard is placed inside the cartoon boxes, which are then kept under cold storage with a temperature of 0.5°C and RH maintained at 85-90 per cent. Usually, refrigerated containers are employed for shipment to overseas market.

Marketing and Export

More than 80 per cent of the total production is consumed as table grapes in India, and more than 70 per cent of the total production is harvested in March-April, but the cold storage facilities are inadequate. Therefore, market gluts and fall of prices of grapes in March-April are common. Approximately, 2.5 per cent (22,000 t) of fresh grapes are exported to the Middle East and European countries. The rest of the produce is marketed within the country. Grapes are exported through three different agencies *viz.*, Grower Exporters, Growers' Cooperatives and the Trader exporters. These agencies have established their own facilities for pre-cooling and cold storage in the vicinity of major production sites.

There is also a need for diversification of the uses of grapes. Currently more than 80 per cent of the produce is used for table purposes. The major bulk of the produce is harvested in March-April, but as cold storage facilities are currently inadequate, there are common occurance of heavy market gluts, that leads to huge wastage of products. Diversification of uses as various processed products such as; raisins, wine, juice *etc.* and export of table grapes can ease the marketing problems. Maintenance of quality of table grapes by crop regulation is the priority consideration to increase exports.

Government Policies

The Government of India is supporting the grape industry of the country in the following ways:

a) Encourage and support the farmers for establishing the vineyards and installing drip irrigation systems by providing soft loans and subsidies.
b) Provide research support to sustain the productivity of grapes under adverse situations.
c) Promote and support the export of fresh grapes by training the growers and providing soft loans and subsidies for pre-cooling and cold storage facilities.

Research on grapes is carried out by the Indian Council for Agricultural Research (ICAR) Institutes and State Agricultural Universities at different centres under the All India Coordinated Research Project on Grapes. The National Research Centre for Grapes (ICAR) located at Pune, Maharashtra is the focal point for conducting and coordinating the research activities on grapes throughout the country.

Conclusion

Though grape is one of the fruit crops with very high productivity, the area under vineyard seems not expanding as much as it can owing to the high initial cost of establishing the vineyards and high recurring cost of production. The risk of losing a crop due to fluctuations in weather under changing climate scenario is also very high. Since the highest productivity in grapes across the world has been achieved in India, proper measures are needed to extend grape cultivation to newer areas. Various abiotic stresses including drought, salinity, *etc.* are the main challenges for which suitable rootstocks are to be identified and commercialized. For the survival of the grape industry in India, the produce should be quality and cost competitive. Hence, future efforts are to be carried forward in this direction.

Chapter 5

Microbes: A Way Towards Organic Farming

R.K. Sahu, Vinod Kumar, Shish Ram Jakhar, Rokozeno Chalie-u and N.G. Mitra

Department of Soil Science and Agricultural Chemistry, J.N. Krishi Vishwa Vidyalaya, Jabalpur – 482 004, Madhya Pradesh

Introduction

How Much Fertilizer Actually Goes to the Plant?

Did you know that only 40 to 60 per cent of the fertilizer we apply actually goes to the plant, the remaining is lost to run off into our waterways, volatilization to the air or is tied up in the soil. This is why soil health is such an imperative piece of plant health. Functional soil is a soil embedded with organic matter and soil microbes that work together to hold onto nutrients in the soil and convert nutrients locked in the soil.

Beneficial soil microbes form symbiotic relationships with the plant. In fact, the plant will exert as much as 30-40 per cent of its energy to the root zone to make food for microbes. In return those microbes not only protect the plant from stress, but also feed the plant by converting and holding nutrients in the soil.

What are the Different Types of Soil Microbes?

There are five different types of soil microbes: bacteria, actinomycetes, fungi, protozoa and nematodes. Each of these microbe types has a different job to boost soil and plant health.

Bacteria

Bacteria is the crucial workforce of soils. They are the final stage of breaking down nutrients and releasing them to the root zone for the plant. In fact, the Food and Agriculture Organization once said "Bacteria may well be the most valuable of life forms in the soil."

Actinomycetes

Actinomycetes were once classified as fungi, and act similarly in the soil. However, some actinomycetes are predators and will harm the plant while others living in the soil can act as antibiotics for the plant.

Fungi

Like bacteria, fungi also lives in the root zone and helps make nutrients available to plants. For example, *Mycorrhizae* is a fungi that facilitate water and nutrient uptake by the roots and plants to provide sugars, amino acids and other nutrients.

Protozoa

Protozoa are larger microbes that love to consume and be surrounded by bacteria. In fact, nutrients that are eaten by bacteria are released when protozoa in turn eat the bacteria.

Nematodes

Nematodes are microscopic worms that live around or inside the plant. Some nematodes are predators while others are beneficial, eating pathogenic nematodes and secreting nutrients to the plant.

Distribution of Microorganisms within the Soil Profile

The abundance and biomass of most soil organisms is highest in the top 10 cm of soil and declines with depth in parallel with organic matter content and prey availability. Approximately 65 per cent of the total microbial biomass is found in the top 25 cm of the soil profile. Below that depth, microbial densities typically decline by one to three orders of magnitude. Abundances of G- bacteria, fungi, and protozoa are typically highest at the soil surface, whereas G+ bacteria, Actinobacteria, and archaea tend to increase in proportional abundance with increasing depth (Fierer *et al.*, 2003). In forest soils, the litter layer is dominated by fungi, with fungal biomass being up to three times higher than that of bacterial biomass (Baldrian *et al.*, 2012). The relative importance of bacteria increases in the organic and mineral soil horizons. Hyphal density of and root colonization by mycorrhzial fungi decrease substantially below 20 cm. Microbial grazers (*e.g.*, protozoa, collembola) also decrease with depth, often more rapidly than either their bacterial or fungal prey. Collembolan (hexapod group, 6 abdominal segments; segment 1 ventral tube has glue peg/piston with adhesive and water balance properties) numbers peak at 1-5 cm below the soil surface and drop to almost none below 10 cm.

Although generally low, the numbers and activities of soil organisms at depth vary spatially depending on gradients in texture, pH, temperature, water availability, and organic matter content.

Top soil is rich in soil organic matter – dark spongy material formed from decomposed plant and animal tissue. Soil organic matter is critically important: It helps soils hold onto water and nutrients and supports soil microbes that recycle nutrients. Loss of soil organic matter has made many farms increasingly reliant on fertilizers, pesticides and herbicides.

Much recent research has focused on adding organic material back to soils to restore them. This is an important strategy, but it is evidently believed that we also should aim to enhance the microbes that are responsible for soil formation. It was demonstrated in a 2015 study (published by Kaitlyn, 2018) that adding efficient microbes to soils can enhance the percentage of plant carbon that is transformed into soil. New research suggests that by fostering an efficient and active soil microbiome, we can accelerate soil regeneration far beyond typical rates seen in nature.

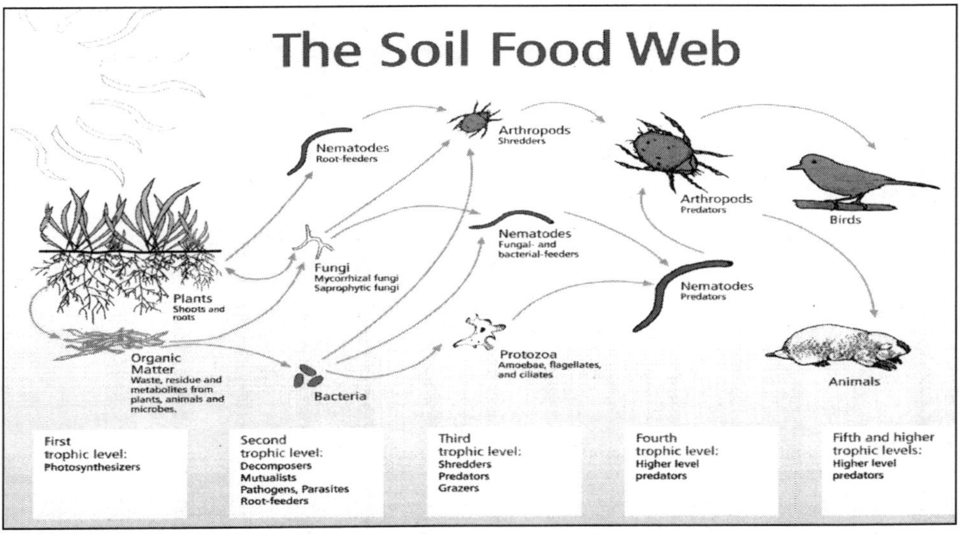

Figure 5.1: Microbes Perform Critical Functions in Soil Food Webs, such as Decomposing Organic Materials, Cycling Nutrients and Improving Soil Structure.

Natural soils are thriving with life. They contain an incredible diversity of microscopic bacteria, actinomycetes, fungi, viruses and other organisms as well. One g of soil can contain $10^7 - 10^9$ prokaryotes with 10000 of different species.

These microbes interact closely with each other, forming complex networks. They communicate with chemical signals (quorum sensing or quenching). They work together to break down complex organic materials, including dead plants and animals. They often work in teams to complete biochemical processes, such as transforming nitrogen from an inert gas to plant-usable forms, other nutrients like P, K, S, and micronutrients, and recycling them from dead plant materials back into dissolved forms.

In healthy soils, organic matter is protected from decomposition inside clumps of soil called aggregates. But tilling crushes aggregates, unlocking their carbon and allowing microbes and soil fauna to attack it. This creates a temporary feast for soil microbes, but eventually they deplete their food supply and die off. Without a healthy microbial community, nutrients are no longer recycled, opportunistic pests can invade and farmers rely increasingly on chemicals to replace biological soil functions.

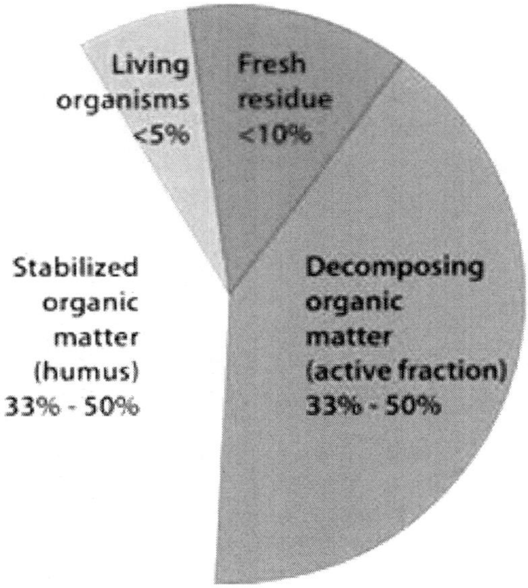

Figure 5.2: Components of Soil Organic Matter.

Reviving Agricultural Soils

Soil degradation is a critical problem because it threatens our ability to produce enough healthy food for a growing human population and contributes to climate change. In response, large companies, nonprofits, scientists and government agencies are working together to restore soil health.

The first step to improving soil health is to stop the bleeding. Instead of leaving fields barren in between crops, which leads to erosion, farmers are increasingly planting cover crops such as rye grass, oats and alfalfa. They also are replacing intensive tilling with no-till practices to prevent the breakdown of soil structure.

Soil organic matter contains over 50 per cent carbon. Globally, soils contain more carbon than plants and the atmosphere combined. Losing carbon-rich organic matter from soils releases carbon dioxide, a greenhouse gas, which can accelerate climate warming. But by regenerating our soils, we can sequester more carbon underground and slow climate warming.

In addition to protecting soil, cover crops take carbon out of the atmosphere as they grow and funnel it into the soil. Unlike cash crops that are harvested and removed from the soil, cover crops are left to decompose and contribute to soil formation.

Increasing the supply of plant carbon in this way is an important first step in rebuilding soil carbon. But new research suggests that it may be insufficient.

A New Paradigm of Soil Formation

We used to think that soil organic matter was formed from leftover bits of plants that were difficult to degrade. Over time, we thought that these plant particles became chemically transformed into what was called humus – dark, long-lasting material left over when dead plants and animals decay. This view suggested that the key to building soils was getting a lot of dead plant material into the ground.

Recently, however, technological advances have transformed our understanding of soil formation. There is now strong evidence that that the most persistent forms of soil carbon are formed primarily from dead microbial bodies rather than from leftover plant parts. The vast majority of old soil carbon appears to have undergone microbial decomposition. While plants are the original source of carbon for soils, microbes control its fate by using it as food, thus ensuring that at least some of it will remain in the soil.

Feeding the Soil by Feeding Beneficial Microbes

Microbes can take a simple compound like sugar and transform it into the thousands of complex molecules found in soils. When microbes break plant matter down, they use some of the material they consume for building new biomass – that is, to fuel their own growth – and exhale the rest as carbon dioxide. The efficiency with which they create new biomass varies widely. Some microbes are like weeds: They grow quickly in food-rich environments, but are sloppy eaters and waste much of what they consume. Others are slow-growing but hardy, waste little and are able to survive times of starvation or stress.

To maximize the proportion of plant carbon that is transformed into soil organic matter, we should aim to support and enhance soil microbiomes that quickly and efficiently transform dead plant materials into soil organic matter. Healthy soils should also contain microbiomes that help prevent disease, cycle nutrients and help reduce plant stress.

My research group is now bioprospecting for groups of microbes that are especially efficient at forming new soil and recycling nutrients that help enhance soil health- making soils more healthy. These will make it possible to grow more food with fewer inputs, which will make farming more profitable and protect our air and water.

A more holistic concept of farming in terms of agro-ecology has begun to challenge the traditional reduction approach to the study of agriculture. As a result we have become more aware of the importance of microorganisms in such processes as soil formation, plant nutrition, and the suppression of plant pathogens, pests and weeds. Moreover, it has become clear that many farming practices – intensive tillage, pesticide use, fertiliser use and mono-cropping in particular – are directly or indirectly harmful to soil microbes and therefore to the processes mediated by them. It is of concern that since much industrialized agriculture now relies heavily on manufactured inputs rather than using natural resources, the sustainability of agriculture is threatened and a move towards resource conserving agriculture is essential. Farming practices must be designed to optimize soil microbial life as part of a wider strategy aimed at conserving and replenishing natural resources. The practical application of this technology and the role of naturally occurring populations of microorganisms in agro-ecosystems form the basis of this article.

Soil microbial biomass is composed of eukaryotic (fungi, yeasts, protozoa and algae), and prokaryotic (eubacteria, actinomycetes and archaea) organisms, whose populations vary from soil to soil (Shannon *et al.,* 2002). Soil microbial biomass is dictated by a variety of soil and environmental parameters including soil texture and structure, pH, air/moisture content, and soil temperature (Campbell *et al.,* 1999). Now it is well obvious that one of the prime factors that determine soil microbial status is the type and amount of organic material that enters the soil ecosystem. The vast majority of soil microorganisms are heterotrophic and require organic materials as both carbon and energy sources (Shannon *et al.,* 2002). Management practices in particular, the manipulation of the quality and quantity of organic inputs might, therefore, be predicted to modify soil microbial populations, the soil food web (trophic levels), and the biological processes involved in nutrient transformation (Stockdale *et al.,* 2002).

Microorganisms play a fundamental role in soil creation and stability through the binding of soil aggregates by hyphae and by the secretion of exudates (Andrade *et al.,* 1998). There are many examples of the ways in which soil beneficial microbial systems (microbial inoculants) help ecosystem health and stability.

Microbial inoculants based on these microorganisms can be divided into different categories depending on their use.

- ✰ Biofertilizers are substances which contain microorganisms in living/dormant cells which, when applied to seed, plant surfaces, or soil, colonize the rhizosphere or the interior of the plant, and promote growth by increasing the supply or availability of nutrients to the host plant.
- ✰ Three major groups of microorganisms are considered beneficial to plant nutrition: nitrogen-fixing bacteria (diazotrophs), P and/or other nutrients solubilizing/mobilizing microorganisms including arbuscular mycorrhizal fungi (AMF- *Gigaspora, Funneliformis* or *Rhizophagus* (*Glomus*), and *Laccaria*), plant growth-promoting rhizobacteria (PGPR).
- ✰ Another category of PGPM-containing products is that of phytostimulators/plant biostimulants which are generally containing auxin-producing bacteria, inducing root elongation.

The widely accepted definition of plant biostimulants (du Jardin, 2015) is: substance(s) and/or micro-organisms whose function when applied to plants or the soil rhizosphere stimulates the natural processes to enhance/benefit nutrient uptake and efficiency, tolerance to abiotic stress, and crop quality. Typically, biostimulants do not have a direct action against pests.

Classification of Biofertilizers

Biofertilizers

- **N fixers**
  - **Bacteria**: *Rhizobium*, *Azotobacter*, *Azospirillum*, *Mycobacterium*, *Bacillus*
  - **BGA**: *Anabaena*, *Anabaenopsis*, *Nostoc*, *Tolypothrix*
  - **Azolla**: *Azolla filiculoides*, *Azolla rubra*
- **PSM**: *B. megatherium*, *Serratia marcescens*
- **P mobilizer**: *G. phasiculatus*
- **Other nutrient solubilizers**: *T. thiooxidans* (S), *F. aurentia* (K), *T. ferrooxydans* (Fe), *P. auriginosa* (Zn)
- **PGPR**: *P. fluorescens*, *T. viride*

Symbiotic N_2 Fixation

A Short List of Symbiotic Species and their Corresponding Hosts

Symbiotic Species	*Host Plants*	*Cross Inoculation Group*
Bacteria: *Rhizobium*		
Bradyrhizobium japanicum	*Glycine max* (soybean)	Soybean group
Rhizobium fredii	*Glycine max* (soybean)	
R. leguminosarum	*Pisum sativum* (peas)	Pea group
	Vicia faba (broad bean)	
	Lathyrus (Khesari)	
	Lens	
R. phaseoli	*Phaseolus vulgaris* (common bean)	Bean group
R. meliloti	*Medicago sativa* (alfalfa-hathi ghas)	Alfalfa group
	Melilotus sp. (sweet clovers)	
	Trigonella foenumgraecum (fenugreek-methi)	
R. trifolii	*Trifolium* sp. (clovers)	Clover group
R. lupini	*Lupinus, Ornithopus*	Lupin group
"Cowpea *rhizobia*" group or *Rhizobium* sp.	*Vigna unguiculata* (cowpea),	Cowpea group
	Arachis hypogaea (peanut),	
	Vigna subterranea (Bambara groundnut)	
	Leucaena sp., *Albizia* sp.	
	Crotalaria	
	Pueraria	
Azorhizobium caulinodans	*Sesbania* sp. *Sesbania rostrata* (stem nodulating)	

Major Groups of *Rhizobia*

Genus	*Characteristics*	*Closest Relatives*
Rhizobium	Grow fast on culture media, and genes of nitrogen fixation and nodulation reside on Sym-plasmid except for *R. loti*	*Agrobacterium* (Crown-gall inducing bacteria)
Bradyrhizobium	Grow slowly on culture media, and genes of nitrogen fixation and nodulation reside on chromosome	*Rhodopseudomonas palustris* (photo-synthetic bacteria)
Azorhizobium	Grow slowly on culture media, and genes of nitrogen fixation and nodulation reside on chromosome. Capable of free-living nitrogen fixation. Isolated from the stem nodules of *Sesbania rostrata*	*Aquabacter spiritensis*

In addition to these bacteria, *Frankia, Rhizobium* sp. and cyanobacteria undergo symbiosis by getting established inside the plant tissues and may or may not develop special symbiotic structures. The classical examples of symbiotic association developed by *Rhizobium* sp. are found in about 13,000 or more leguminous plants, both cultivated and non-cultivated herbs, shrub and trees. Among leguminosae, the largest number of plants is in papilionoidae. Moreover, species of *Anabaena, Nostoc, Tolypothrix, etc.* develop symbiotic association with fungi (symbiotic structure is lichen), bryophytes, pteridophytes, gymnosperms and angiosperms and fulfill the requirement of nitrogen deficiency. However, actinorrhizic nodules are developed by *Frankia,* a member of Actinomycetes, on roots of about 170 species of woody dicot non-leguminous plants from 15 genera belonging to temperate trees like *Alnus, Myrica, Casuarina, etc.* (Goodfellow and Williams, 1983). Nodules; are of two types: (i) *Alnus* type where nodules show dichotomous branching to form a corraloid root, and (ii) *Myrica/Casurina* type in which case the apex of each nodule produces a normal but negative geotropic root. The function of nodules is to facilitate gas diffusion to the nitrogen fixing endophyte in the nodule under low O_2 tension (Becking, 1982).

Non-Symbiotic Nitrogen Fixation

Microorganisms which pass independent life and fix atmospheric nitrogen are known as free living diazotrophs. There are two groups of such microorganisms: bacteria and cyanobacteria (blue-green algae). Based on the mode of nutrition (carbon, nitrogen and oxygen and requirement of reducing groups) bacteria are divided into (i) aerobic bacteria (*Azonomas, Azotobacter, Beijerinckia, Mycobacterium, Methylomonas*), (ii) facultative anaerobic bacteria (*Bacillus, Enterobacter, Klebsiella, etc.*), (iii) anaerobic bacteria (*Clostridium, Desulfovibrio, etc.*) and (iv) photosynthetic bacteria (*Rhodomicrobium, Rhodopseudomonas, Rhodospirillum, Chromatium, Chlorobium, etc.*).

Among cyanobacteria, both heterocystous and non-heterocystous forms fix atmospheric nitrogen, for example, *Anabaena, Anabaenopsis, Aulosira, Calothrix, Cylindrospermum, Gloeocapsa, Lyngbya, Nostoc, Oscillatoria, Plectonema, Scytonema, Stegonema, Tolypothrix, Trichodesmium, etc.* (Alexander, 1977).

P Solubilization/Mobilization

Strategies for mobilizing soil P, namely acidification of the soil, the release of organic acid anions (*e.g.* citrate, oxalate, gluconate), protonation and the release of enzymes acid and alkaline phosphatases and phytase (Yadav and Tarafdar 2007; Xiao *et al.*, 2009), chelation of cations and competing with phosphate for adsorption sites in the soil

K Solubilization

The mechanisms for KSMs to solubilization of K are by: (i) lowering the pH, or (ii) by enhancing chelation of the cations bound to K, and (iii) acidolysis of the surrounding area of microorganism. The lowering in pH of the medium suggests the release of organic acids and protons by the K-solubilizing microorganisms (Zarjani *et al.*, 2013; Parmar and Sindhu 2013; Uroz *et al.*, 2009). Such acidolysis by organic acids produced by the rhizospheric microorganisms can either directly dissolve the mineral K as a result of slow releases of exchangeable K, readily available exchangeable K or can chelate both Si and Al ions associated with K minerals (Romheld and Kirkby 2010). Thus, the synthesis and discharge of organic acids (*e.g.* tarterate) by the microorganisms into the surrounding environment acidify the microbe's cells and their surrounding environment that ultimately lead to the release of K ions from the mineral K by protonation and acidification (Goldstein 1994). Of the different organic acids involved in the solubilization of insoluble K, tarteric, succinic, citric, gluconic, α-ketogluconic and oxalic acids are the most prominent acids released by microbial strains

Plant Growth-Promoting Rhizobacteria (PGPR)

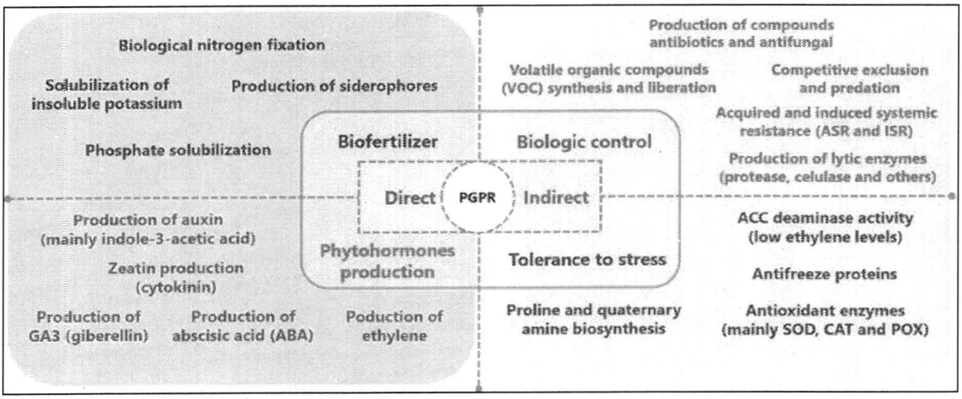

Figure 5.3: Direct and Indirect Mechanisms Mediated by Plant Growth-Promoting Rhizobacteria (PGPR) with Beneficial Effects on Host Plants (Chauhan *et al.*, 2015; Pii *et al.*, 2015).

Mechanisms of Action

Direct Mechanisms

1. Biofertlizer: Biological nitrogen fixation, P and K solubilization,
2. Siderophores production, and
3. Phytohormones production (IAA, Ctokinin, GA, ABA, Ethylene)

Indirect Mechanisms

1. Biological control
   Antibiotics, bacteriocins and antifungal
   Volatile organic compounds
   Competitive exclusion and predation
   Acquired and induced sytamic resistance (ASR and ISR)
   Lytic enzymes (protease, cellulose and others)
2. Bioremediation (Plant-Microbe-Metal Interactions)
   Metal mobilization
   Metal immobilization
   Metal transformation
   Metal transport and distribution
3. Tolerance to stress
   ACC deaminase activity
   Antifreeze proteins
   Proline and quaternary amines (osmolytes)
   Antioxidant enzymes (SOD, CAT and POX)

Mycorrhizal Microorganisms

Mycorrhizal fungi can be broadly categorised into two groups: vesicular arbuscular mycorrhizas (VAM or AM) and ecto-mycorrhizas (EM), which vary widely in structure and function (Harrier, 2001). The former are microscopic obligate symbionts that can only be cultured with a host plant and constitute the largest proportion of fungal biomass in the soil. VAM produce structures known as arbuscules and vesicles within the root cell that are involved in the nutrient transfer process. This symbiosis confers benefits directly to the host plants through the acquisition of phosphate and other mineral nutrients from the soil by the fungus while the fungus receives a carbon source (primarily glucose) from the host and store in the fungal mycelia as trehalose. In addition, the symbiosis may also enhance the plants resistance to biotic and abiotic stresses. Ectomycorrhizas (EM), regularly produce large above-ground fruiting bodies (like mushroom and toadstool) as well as a hyphal net around the plant root. Vesicles and arbuscules are absent and hyphal penetration of the root is limited. EM fungi are amenable to axenic culture. Potential inoculums can also be produced in the field from mycelium.

Mycorrhizal fungi are also believed to control root pathogens by one or more of: (1) improving nutrient acquisition by host plant; (2) excluding pathogens at root infection sites and within the rhizosphere; (3) inducing anatomical and structural changes in the root thereby creating physical barriers to pathogen entry; (4) producing substances antagonistic to root pathogens; and (5) activation of plant defence mechanisms. The role of mycorrhizal fungi, as well as other microorganisms, in stimulating induced resistance to soil-borne plant pathogens in a range of crops has received much attention over the past 25 years (Azcon-Aguilar and Barea, 1996) and it is worth considering its significance to organic farming. Interactions between plants, mycorrhizal fungi and soil bacteria are not random but appear to occur in an ordered manner between compatible species. The presence in the soil of given mycorrhizal species can influence the persistence and activities of native bacteria and *vice versa* (Andrade *et al.*, 1998).

In addition, the positive interaction between composting and VAM inoculation that led to increased growth of tree seedlings was related to the ability of the mycorrhiza to enhance plant nutrient uptake from the compost. The possibilities for exploiting these types of synergism have been investigated in many cases (Srivastava *et al.*, 2002). Heavy metal pollution of soils (along with severe compaction) is common on metalliferous mine spoils and on modern brown-field sites following industrial activities. Polluted soils such as these are slow to support new vegetation and so remain prone to further damage from erosion. One of the ways in which plants benefit from ecto-mycorrhizal infection in nature is through an increased tolerance to heavy metals in the soil. Mycorrhizal fungi confer protection to the plant against uptake of heavy metals by adsorption on hyphae and through various forms of fungal metabolism (Gadd, 1983).

Table 5.1: Application of VAM Inoculum in Agriculture, Horticulture and Forestry

Crop	*VAM Fungus*	*Outcome*	*Source*
Banana	*Acaulospora* sp.	Improved growth and nutritional status	Yano-Melo *et al.*, 1999
Mulberry	*Glomus* sp.	Reduced P requirements, Improved growth	Setua *et al.*, 1999
Douglas fir	*Laccaria bicolor*	60 per cent increased in wood volume	Selosse *et al.*, 2000
Cyclamen	*Glomus* sp.	Increased survival in nursery	Vosatka *et al.*, 1999
Acacia	*Terminalia* sp.	Improved growth and survival	Munro *et al.*, 1999
Douglas fir	*Melanogaster* sp., *Rhizopogon* sp.	Improved growth	Pera 1999
Black spruce	*Paxillus* sp.	Protection against *Cylindrocladium* root rot	Morin *et al.*, 1999

Table 5.2: Plant Probiotic Bacteria Used to Enhance the Quality of Fruit and Horticultural Crops

Bacterial Strain	*Fruit or Horticultural Crop*	*Crop Effects Caused*	*Reference*
Phyllobacterium sp. PEPV15	*Fragraria x ananassa*	Enhancement of vitamin C	Flores *et al.*, 2015
Pseudomonas BA-8, Bacillus OSU-142 and Bacillus M-3	*Fragraria x ananassa*	Enhancement of vitamin C	Pýrlak and Kose 2009
Paenibacillus polymyxa	*Fragraria x ananassa*	Enhancement of vitamin C	Erturk *et al.*, 2012
*Pseudomonas sp.*19Fv1T	*Lycopersicon esculentum*	Enhancement of vitamin C	Bona *et al.*, 2017
Bacillus megaterium and *Bacillus amyloliquefaciens*	*Lycopersicon esculentum*	Enhancement of vitamin C	Shen *et al.*, 2016
Bacillus amyloliquefaciens FZB2 and*B. amyloliquefaciens* FZB42	*Lycopersicon esculentum*	Enhancement of vitamin C	Gul *et al.*, 2008
Pseudomonas sp. 5Vm1k and AMF	*Fragaria x ananassa*	Enhancement of vitamin B and C	Bona *et al.*, 2015
Pseudomonas putida 41, *Azotobacter chrooccum 5, Azospirillum lipoferum* OF and AMF	*Lycopersicon esculentum*	Inreased antioxdant activity	Ordookhani *et al.*, 2010
Bacillus licheniformis	*Lycopersicon esculentum*	Improved total flavonoids content	Ochoa *et al.*, 2016
Pseudomonas putida 41, *Azotobacter chroococcum* 5, *Azospirillum lipoferum* OF	*Ocimum basilicum*	Inreased antioxdant activity	Ordookhani 2011
Pseudomonas sp., *Bacillus lentus* and *Azospirillum brasilens.*	*Ocimum basilicum*	Inreased antioxdant activity and chlorophyll leaf content	Heidari and and Golpayegani 2012
Rhizobium leguminosarum PETP01	*Capsicum annuum*	Inreased antioxdant activity	Silva *et al.*, 2014
Pseudomonas fluorescens N21.4	*Rubus sp.*	Inreased flavonoids concentration	García *et al.*, 2015
Rhizobium sp. PEPV12	*Spinacia oleracea.*	Inreased chlorophyll content	Jiménez *et al.*, 2016
Paenibacillus polymyxa RC14	*Brassica oleracea var capitata cv Yalova 1.*	N,P,K,S,Fe and Cu content increase	Yildirim *et al.*, 2015
Bacillus subtilis BA-142, *Bacillus megaeorium*- GCsub group A.MFD- 2, *Acinetobacter baumannii* CD-1 and *Pantoea agglomerans* FF	*Lycopersicon esculentum L. and Cucumis sativus L.*	N,P,Mg, Ca, Na, K, Cu, Mn,Fe and Zn content increase in both fruit	Dursun *et al.*, 2010
Bacillus subtilis EY2, *Bacillus atrophaeus* EY6, *Bacillus spharicus* GC subgroup B EY30, *Staphylococcus kloosii*, and *Kocuriaery thromyxa* EY43	*Fragaria x ananassa*	N,P,Mg, Ca, Mg, S, Cu, Mn and Fe	Karlidag *et al.*, 2013

Bacterial Strain	Fruit or Horticultural Crop	Crop Effects Caused	Reference
Pseudomonas fluorescens and *Bradyrhizobium sp.*	*Origanum majorana*	Increased the amount of essential oils	Banchio *et al.*, 2008
Bacillus megaeorium, Pantoea agglomerans and *Bacillus subtilis*	*Brassica oleracea*	Inreased chlorophyll plant content	Turan *et al.*, 2014
Pseudomonas fluorescens N21.4	*Rubus sp. var Lochness*	Increased and stabilize total flavonoid content during different season	Ramos *et al.*, 2014

Table 5.3: Benefits in Use of Biofertilizers

Sl.No.	Biofertilizer	Crop Inoculated	Nutrient Saved (kg ha^{-1})	Crop Yield Increased (Per cent)
1	*Rhizobium*	Pulses and oil seeds	20-30 N	10-20
2	*Azospirillum*	Cereals and Vegetables	20-30 N	15-20
3	*Azotobacter*	All crops (non-leguminous)	10-15 N	7-12
4	*Azolla*	Paddy	25-50 N	10-15
5	Blue green algae	Paddy	20-30 N	10-15
6	VA mycorrhiza	All crops	-	15-20
7	P-solubilizing bacteria (*Bacillus, Pseudomonus, etc.*)	Legumes, cereals and vegetables	15-30 P_2O_5	15-20

Table 5.4: Jawahar Biofertilizers Produced in MRPC, JNKVV

Sl.No.	Biofertilizers (type/group)	Crop
1	Jawahar legume nitro (*Rhizobium*)	Legumes
2	Jawahar non-legume/cereal nitro (*Azobacter/Asosprillum*)	Cereals
3	Jawahar BGA	Paddy
4	Jawahar cash nitro (*Acetobacter*)	Cash crops
5	Jawahar phospho (PSB)	All crops
6	Jawahar mycorrhiza	All crops
7	Jawahar potash (KSB)	All crops
8	Jawahar micro (ZSB)	All crops
9	Jawahar PGPR (*Pseudomonas*)	All crops
10	Jawahar plant stimulator (Biofertisol)	All crops
11	Jawahar protector (*Trichoderma*)	All crops
12	Jawahar biodigester (consortia)	Waste materials
13	Jawahar enriched bio-organic (MOM)	All crops

Methods of Application

Powder Culture

Seed treatment: 15-20 g culture/kg seed

Soil treatment: 3-4 kg culture/acre, mix the culture with 50 kg FYM/compost/ soil and apply

Liquid Culture

Seed treatment: 10 ml culture/kg seed

Soil treatment: 2 litre culture/acre, mix the culture with 50 kg FYM/compost/ soil and apply

Drip/foliar application: 1.5-2.0 litre culture/acre, mix the culture with sufficient quantity of water and apply

Organic Agriculture and Microorganisms

Soil management practices that change the soil carbon content also influence the size and structure of the soil biomass. Application of organic amendments and/or cover crops increases carbon availability to microorganisms. Schjonning *et al.* (2002) found that, after more than 40 years, microbial biomass C was higher in organic than in conventionally managed dairy farm soils. Carbon released from crop residues contributes to increasing soil microbial activity and so increases the likelihood of competition effects in the soil. The placement of the residue in soil can lead to the displacement of a pathogen from its preferred niche, diminishing the pathogen's ability to survive (Bailey and Lazarovits, 2003). Therefore, soil microbial biomass changes as a consequence of switching to organic land management (Shannon *et al.*, 2002). Application of organic matter and all treatments that increase the total microbial activity in the soil might enhance the general suppression of pathogens by increasing competition for nutrients. Crop rotation may also yield microbial benefits beyond those normally associated with pathogen host range and saprophytic survival. Rotation is most successful in limiting the impact of biotrophic pathogens that require living host tissues, or those pathogens with low saprophytic survival capability (Bailey and Duczek, 1996).

Agricultural practices can have major short- or long-term impacts on mycorrhizal fungi, as well as on other soil microorganisms. Scullion *et al.* (1998) studied the effectiveness of VAM spores from organically and conventionally managed soils and found that white clover only benefited from mycorrhizal infection in a low-fertility (organically managed) soil. Furthermore, inocula from organic soils were more effective in both achieving mycorrhizal infection and in allowing more efficient P uptake by the crop. Application of composts may enhance beneficial soil microorganisms. As the active microbial biomass increases, the capacity to utilize carbon, nutrients and energy in the soil is increased and thus these resources might be limited for soil-borne pathogens (Sullivan, 2001).

Conclusion

It is clear that soil microorganisms play a fundamental role in soil health and sustainability. Microbial population density and diversity are affected by the level of organic matter indirectly by providing energy for soil microorganisms, improving soil structure and stability, soil moisture and plant nutrient availability and by preventing soil-borne disease incidence. Crop and soil management practices designed to support soil microorganisms *i.e.* minimum tillage, cover cropping, composting, eliminating pesticide and fertilizer use, and maintaining biodiversity, will inevitably lead to soil improvements and stability. The need for replenishing organic matter in the soil after harvesting is thus of great importance. In ecological farming systems, organic matter is maintained by mixed farming, rotations, recycling, compost and farm yard and green manures and bought-in organic sources. Including legumes and cover crops in the rotation should help balanced soil fertility, disease management, and avoid leaching nutrients and soil pollution. Crop and soil management practices designed to support soil microorganisms have been shown to lead to improvements in soil structure and stability. Although the great diversity of soil microorganisms is now well appreciated, studies over the past twenty years have established that a number of other soil organisms, apart from mycorrhizal fungi, not only play an active role in suppressing pathogens but also mediate the activity of a range of beneficial organisms. Apart from farming in ways that support indigenous microbes, the use of effective exotic cultures of beneficial microorganisms as soil and compost amendments, root dips, feed additives or sprays in order to supplement or reinforce the indigenous micro-flora are being considered.

REFERENCES

Alexander M 1977. Introduction to Soil Microbiology, pp. 467. Academic Press, New York.

Andrade G, Linderman RG and Bethlenfalvay GJ. 1998. Bacterial associations with the mycorrhizosphere and hyphosphere of the arbuscular mycorrhizal fungus Glomusmosseae. *Plant Soil*. 202: 79-87.

Andrade G, Mihara KL, Linderman RG and Bethlenfalvay GJ. 1998. Soil aggregation status and rhizobacteria in the mycorrhizosphere. *Plant Soil*. 202: 89-96.

Azcon-Aguilar C and Barea JM. 1996, Arbuscularmycorrhizas and biological control of soil-borne plant pathogens- an overview of mechanisms involved, *Mycorrhiza* 6: 457-464.

Bailey KL and Duczek LJ. 1996. Managing cereal diseases under reduced tillage. *Can. J. Plant Pathol*. 18: 159–67.

Bailey KL and Lazarovits G. 2003. Suppressing soil-borne diseases with residue management and organic amendments. *Soil Till Res*. 72: 169-80.

Baldrian P, Kolar¡´ýk, M, S¡tursova´ M, Kopecky´ J, Vala´s¡kova´ V, Ve¡trovsky´, T, Zifc¡a´kova´, L, Snajdr, J, Rý´dl, J, Vlc¡ek, C¡ and Vor¡´ýs¡kova´, J. 2012. Active and total microbial communities in forest soil are largely different and highly stratified during decomposition. *ISME J*. 6: 248–258.

Banchio E, Bogino PC, Zygadlo J and Giordano Wl. 2008. Plant growth promoting rhizobacteria improve growth and essential oil yield in *Origanum majorana* L. *Biochem Syst Ecol*.36: 766-771.

Becking JH. 1982. Nitrogen fixation in nodulated plants other than legumes. In: Advances in Agricultural Microbiology. N. S. Subba Rao ed, pp. 89-110. Butterworth Scientific, London.

Bona E, Cantamessa S, Massa N, Manassero P, Marsano F, Copetta A, Lingua G, D'Agostino, G, Gamalero E and Berta G. 2017. Arbuscularmycorrhizal fungi and plant growth-promoting pseudomonads improve yield, quality and nutritional value of tomato: a field study. *Mycorrhiza*. 27: 1-11.

Bona, E., Lingua, G., Manassero, P., Cantamessa, S., Marsano, F., Todeschini, V., Copetta, A., D'Agostino, G., Massa, N., Avidano, L., Gamalero, E. and Berta, G. 2015. AM fungi and PGP pseudomonads increase flowering, fruit production, and vitamin content in strawberry grown at low nitrogen and phosphorus levels. *Mycorrhiza* 25: 181-193.

Campbell CA, Lafond GP, Biederbeck VO, Wen NG, Schoenau J and Hahn D. 1999. Seasonal trends in soil biochemical attributes: Effects of crop management on a black chernozem. *Can. J. Soil Sci*. 79: 85-97.

Chauhan H, Bagyaraj DJ, Selvakumar G and Sundaram SP 2015. Novel plant growth promoting rhizobacteria -prospects and potential. *Appl Soil Ecol* 95: 38–53.

du Jardin P. 2015. Plant biostimulants: definition, concept, main categories and regulation. *Sci. Hortic*. 196: 3–14.

Dursun A, Ekinci M and Donmez MF 2010. Effects of foliar application of plant growth promoting bacterium on chemical contents, yield and growth of tomato (*Lycopersicon esculentum* L.) and cucumber (*Cucumis sativus* L.). *Pak J Bot* 42: 3349-3356.

Erturk Y, Ercisli S and Cakmakci R. 2012. Yield and growth response of strawberry to plant growth-promoting rhizobacteria inoculation. *J Plant Nutr*. 35: 817-826.

Fierer N, Schimel JP and Holden PA 2003. Variations in microbial community composition through two soil depth profiles. *Soil Biol. Biochem*. 35: 167–176.

Flores-Félix, JD, Silva LR, Rivera LP, Marcos-Garcia M, Garcia-Fraile P, Martinez-Molina, E, Mateos, PF., Velasquez, E, Andrade, P and Rivas R. 2015. Plants probiotics as a tool to produce highly functional fruits: the case of *Phyllobacterium* and vitamin C in strawberries. *PLoS One* 10: e0122281.

Gadd GM. 1983. Interactions of fungi with toxic metals. *New Phytol*. 124: 25-60.

García-Seco D, Zhang Y, Gutierrez-Mañero FJ, Martin C and Ramos-Solano B. 2015. Application of *Pseudomonas fluorescens* to blackberry under field conditions improves fruit quality by modifying flavonoid metabolism. *PloS One* 10: e0142639.

Goldstein AH 1994. Involvement of the quinoprotein glucose dehydrogenase in the solubilisation of exogenous phosphates by gram negative bacteria. Phosphate in microorganisms: cellular and molecular biology. ASM Press, Washington, DC, pp197-203.

Goodfellow M and Williams ST 1983. Ecology of actinomycetes. *Annual Review Microbiology*. 37: 189-216.

Gül A, Kidoglu F and Tüzel Y 2008. Effects of nutrition and *Bacillus amyloliquefaciens*on tomato (*Solanum lycopersicum* L.) growing in perlite. *Span J Agric Res*. 6: 422-429.

Harrier LA 2001. Isolation and sequence analysis of the arbuscularmycorrhizal fungus Glomusmosseae (Nicol. and Gerd.) Gerdemann and Trappe 3-phosphoglycerate kinase (PGK) gene promoter region. DNA Sequence (in press).

Heidari M and Golpayegani A. 2012. Effects of water stress and inoculation with plant growth promoting rhizobacteria (PGPR) on antioxidant status and photosynthetic pigments in basil (*Ocimum basilicum* L.). *J Saudi Soc Agr Sci*.11: 57-61.

Jiménez-Gómez A, Menéndez E, Flores-Félix, JD, Garcia-Fraile, PFM and Rivas R. 2016. Effective Colonization of Spinach Root Surface by Rhizobium, In: *Biological Nitrogen Fixation and Beneficial Plant-Microbe Interaction,* Springer International Publishing, 109-122.

Karlidag H, Yildirim E, Turan M and Pehluvan M. 2013. Plant growth-promoting rhizobacteria mitigate deleterious effects of salt stress on strawberry plants (*Fragaria× ananassa*). *Hort Sci*. 48: 563–567.

Morin C, Samson J and Dessureault M 1999. Protection of black spruce seedlings against Cylindrocladium root rot with ectomycorrhizal fungi. *Can. J. Bot*. 77: 169-74.

Munro RC, Wilson J, Jefwa J and Mbuthia KW. 1999. A low-cost method of mycorrhizal inoculation improves growth of Acacia tortilis seedlings in the nursery. *Forest Ecol. Manage*. 113: 51-6.

Ochoa-Velasco CE, Valadez-Blanco R, Salas-Coronado, R, Sustaita-Rivera F, Hernandez-Carlos B, Garcia-Ortega S and Santos-Sanchez NF 2016. Effect of nitrogen fertilization and *Bacillus licheniformis* biofertilizer addition on the antioxidants compounds and antioxidant activity of greenhouse cultivated tomato fruits (*Solanumlyco persicum* L. var. Sheva). *Sci Hor*. 201: 338-345.

Ordookhani K 2011 Investigation of PGPR on antioxidant activity of essential oil and microelement contents of sweet basil. *Adv Environ Biol*. 5: 1114-1120.

Ordookhani, K., Khavazi, K., Moezzi, A. and Rejali, F. 2010. Influence of PGPR and AMF on antioxidant activity, lycopene and potassium contents in tomato. *Afr J Agri Res*. 5: 1108-1116.

Parmar P and Sindhu SS 2013. Potassium solubilization by rhizosphere bacteria: influence of nutritional and environmental conditions. *J. Microbiol. Res*. 3: 25-31.

Pera J, Alvarez IF, Rincon A and Parlade J. 1999. Field performance in northern Spain of Douglas-fir seedlings inoculated with ectomycorrhizal fungi. *Mycorrhiza*. 9: 77-84.

Pii Y, Mimmo T, Tomasi N, Terzano R, Cesco S and Crecchio C 2015. Microbial interactions in the rhizosphere: beneficial influences of plant growth-promoting rhizobacteria on nutrient acquisition process: a review. *Biol Fert Soils*. 51(4): 403–415.

Pýrlak L and Köse M 2009. Effects of plant growth promoting rhizobacteria on yield and some fruit properties of strawberry. *J Plant Nutr*. 32: 1173-1184.

Ramos-Solano B, Garcia-Villaraco A, Gutierrez-Mañero FJ, Lucas JA, Bonilla A and Garcia-Seco D 2014. Annual changes in bioactive contents and production in field-grown blackberry after inoculation with *Pseudomonas fluorescens*. *Plant Physiol Bioch*. 74: 1-8.

Romheld V and Kirkby EA 2010. Research on potassium in agriculture: needs and prospects. *Plant and Soil*. 335: 155-180.

Schjonning P, Elmholt S, Munkholm LJ and Debosz K 2002. Soil quality aspects of humid sandy loams as influenced by organic and conventional long-term management. *Agric. Ecosyst. Environ*. 88: 195-214.

Scullion J, Eason WR and Scott EP 1998. The effectivity of arbuscularmycorrhizal fungi from high input conventional and organic grassland and grass-arable rotations. *Plant Soil*. 204: 243-54.

Selosse MA, Bouchard D, Martin F and Tacon F 2000. Effect of Laccaria bicolor strains inoculated on Douglas-fir (*Pseudotsugamenziesii*) several years after nursery inoculation. Can. J. Forestry Res. 30: 360-71.

Setua GC, Kar R, Satpathy B, Ghosh JK and Saratchandra B. 1999. Effect of vesicular arbuscularmycorrhiza on growth, leaf yield and phosphorus uptake in mulberry (*Morus alba*) under irrigated, alluvial soil conditions. *Indian J. Agric. Sci*. 69: 833-6.

Shannon D, Sen AM and Johnson DB 2002. A comparative study of the microbiology of soils managed under organic and conventional regimes. *Soil Use Manage*. 18: 274-83.

Shen F, Zhu TB, Teng MJ, Chen Y, Liu MQ, Hu F and Li HX 2016. Effects of interaction between vermicompost and probiotics on soil nronerty, yield and quality of tomato. *YingyongShengtaiXuebao*, 27.

Silva LR, Azevedo J, Pereira MJ, Carro L, Velazquez E, Peix A, Valentao P and Andrade PB. 2014. Inoculation of the nonlegume*Capsicum annuum*(L.) with Rhizobium strains. 1. Effect on bioactive compounds, antioxidant activity, and fruit ripeness. *J Agr Food Chem*., 62: 557-564.

Srivastava AK, Singh S and Marathe RA 2002. Organic citrus: soil fertility and plant nutrition. *J. Sustainable Agric*. 19: 5-29.

Stockdale EA, Shepherd MA, Fortune S and Cuttle SP. 2002. Soil fertility in organic farming systems – fundamentally different? *Soil Use Manage*. 18: 301-8.

Sullivan, P. 2001. Sustainable management of soil-born plant diseases. ATTRA, USDA's Rural Business Cooperative Service. www. attra. org.

Turan, M, Ekinci, M and Yildirim E 2014. Plant growth-promoting rhizobacteria improved growth, nutrient, and hormone content of cabbage (*Brassica oleracea*) seedlings. *Turk J Agr Forest*. 38: 327-333.

Uroz S, Calvaruso C, Turpault MP and Frey-Klett P 2009. Mineral weathering by bacteria: ecologu, actors and mechanisms. *Trends Microbiol*. 17: 378-387.

Vosatka M, Jansa J, Regvar M, Sramek F and Malcova R 1999. Inoculation with mycorrhizal fungi – a feasible biotechnology for horticulture. *Phyton (Austria)*. 39: 219-24.

Xiao CQ, Chi R, He H, Qiu GZ, Wang DZ, Zhang WX 2009. Isolation of phosphate solubilizing fungi from phosphate mines and their effect on wheat seedling growth. *Applied Biochemistry and Biotechnology*. 159: 330-342.

Yadav RS and Tarafdar JC 2003. Phytase and phosphatase producing fungi in arid and semi-arid soils and their efficiency in hydrolyzing different organic P compounds. *Soil Biol. Biochem*. 35(6): 745–751.

Yano-Melo AM, Saggin OJ, Lima-Filho JM, Melo NF and Maia LC. 1999. Effect of arbuscularmycorrhizal fungi on the acclimatization of micropropagated banana plants. *Mycorrhiza*. 9: 119-23.

Yildirim E, Turan M, Ekinci M and Dursun A 2015. Growth and mineral content of cabbage seedlings in response to nitrogen fixing rhizobacteria treatment. *Rom Biotech Lett* 20: 10929-10935.

Zarjani JK, Aliasgharzad N, Oustan S, Emadi M and Ahmadi A 2013. Isolation and characterization of potassium solubilizing bacteria in some Iraniansoils. *Arch Agron Soil Sci*. 59: 1713-1723.

Chapter 6

Crop Modeling: A Quantitative Knowledge about Crop and its Environment

S.K. Pandey[1] and Shreesty Pal[2]

[1]Professor and Head, [2]PhD Scholar,
Department of Horticulture
College of Agriculture JNKVV, Jabalpur, Madhya Pradesh

Studies on crop production have traditionally being carried out applying conventional experience-based agronomic research, in which crop production functions are derived from statistical analysis without referring to the underlying biological or physical principles involved. The application of correlation and regression analysis has provided some qualitative understanding of the variables and their interactions that were involved in cropping systems and has contributed to the progress of agricultural science (Kumar and Chaturevdi, 2009). However, very site specific quantitative information is obtained from this type of analysis. The information acquired can only be reliably applied to other sites where climate, important soil parameters and crop management are similar to those used in developing the original functions. Thus, the quantitative applicability of regression based crop yield models for decision making is severely limited. In addition, due to the unavoidable variability associated with weather, more than 10 years is required to develop statistical relationships that are useful in horticultural decision making. Ref Statistical evidence based on long-term studies generally show that more than 40 per cent of the total variation is usually associated with experimental error (Jame and Cutforth, 1996).

As knowledge is assembled, results achieved from observation change from being qualitative to being quantitative and mathematics can be used as the tool to express biological hypotheses. Advances in computer technology have made it possible to consider the combined influence of several factors in numerous interactions. As a result, for more accurate prediction of crop yield it is possible to quantitatively combine the soil, plant, and climatic systems. Thus, with the availability of inexpensive and powerful computers and with the growing popularity of the application of integrated systems to agricultural practices, a new era of agricultural research and development is emerging (Jones *et al.*, 1993). In crop growth modeling, the integration of various disciplines, such as crop physiology, agrometeorology, soil science and agronomy, in a consistent, quantitative and process-oriented manner, gives current knowledge of plant growth and development.

The available computerized decision support systems allows users to combine technical knowledge accommodated in crop growth models with economic considerations and environmental impact evaluations. DSSAT (Tsuji *et al.*, 1994) is an excellent example of a management tool that enables individual farmers to match the biological requirement of a crop to the physical characteristics of the land to obtain specified objectives. Modeling is a new discipline in the Ghanaian research sector, basic background information regarding application of models in research is not available easily. Model application in our society is hindered by lack of awareness about model structure, possibilities and limitations.

Crop Modeling

Modeling is the use of equations or sets of equations to represent the behaviour of a system. Crop models are computer programmes that mimic the growth and development of crops (USDA, 2007). Model imitates or simulates the behaviour of a real crop by estimating the growth of its components, such as leaves, stems, roots and grains. Thus, a crop growth simulation model carries quantitative information concerning major processes involved in growth and development of the crop apart from forecasting the ultimate state of crop production or harvestable yield.

History of Crop Modeling

The establishment of crop growth simulation models has been a natural advancement of scientific research. Jame (1992) reviewed the history of attempts to quantify the relationships between crop yield and water use from the early work on simple water-balance models in the 1960s to the development of crop growth simulation models in the 1980s. Two decades ago, it was uncertain whether the complex physiological, morphological and physical root processes involved in growth of plants could be catalogued mathematically, except in few controlled environments.

The first attempt to model photosynthetic rates of crop canopies was made in sixties (de Wit, 1965). The results obtained from this model were used among others, to estimate potential food production for some areas of the world and to provide indications for crop management and breeding (de Wit, 1967; Linneman *et al.*, 1979). This was followed by the construction of an Elementary Crop growth Simulator (ELCROS) by de Wit *et al.*, in 1970. This model included the static photosynthesis model and crop respiration was taken as a fixed fraction per day of the biomass, plus an amount proportional to the growth rate. In addition, a functional equilibrium between root and shoot growth was added (Penning de Vries *et al.*, 1974).

The introduction of micrometeorology in the models (Goudriaan, 1977) and quantification of canopy resistance to gas exchanges allowed the models to improve the simulation of transpiration and evolve into the Basic Crop growth Simulator (BACROS) (de Wit and Goudriaan, 1978). To help resource poor farmers in the tropics and sub tropics IBSNAT (International Benchmark Sites Network for Agrotechnology Transfer) began the development of a model in 1982. This was under a contract from the U.S. Agency for International Development to the University of Hawaii at Manoa, USA. IBSNAT was an endeavour to illustrate the efficacy of understanding options through systems analysis and simulation for eventual profit of farm households across the globe. The objectives described for the IBSNAT project by its technical advisory committee were to acknowledge ecosystem processes and mechanisms, synthesize from an understanding of processes and mechanisms, ability to estimate results and facilitate IBSNAT client to apply the predictive capability to control outcomes.

Decision Support System for Agro-Technology Transfer (DSSAT) was the major product of IBSNAT, which at present is utilized as a research and teaching tool. Its role as a research tool is to acquire recommendations analysing crop management and to examine environmental and sustainability issues. The DSSAT products facilitate users to match the biological demands of crops to the physical attributes of land to provide them with management options for improved land use planning. The package comprises of: data base management system for soil, weather, genetic coefficients, and management inputs, Crop simulation models, series of utility and weather generation programs and strategy evaluation program to evaluate options including choice of variety, planting date, plant population density, row spacing, soil type, irrigation, fertilizer application, initial conditions on yields, water stress in the vegetative or reproductive stages of development, and net returns. In fact, DSSAT has the potential to minimize significantly the time and cost of field experimentation requisite for sufficient assessment of new cultivars and new management systems.

Classification of Models

The models can be classified into following groups depending upon the motive for which they are designed:

1. Empirical Models

These models are direct elucidations of observed data and are generally illustrated as regression equations (with one or a few factors). They are used to evaluate the final yield. This approach aims at examining the data, determining an equation or set of equations and fitting them to data. No information about the mechanisms that gives rise to the response is extracted from these models. experiments analysing the response of crop yield to fertilizer application, the relationship between leaf area and leaf size in a given plant species and the relationship between stalk height alone or coupled with stalk number and/or diameter and final yield are some of the examples of these models.

2. Mechanistic Models

The behaviour of the system in terms of lower level attributes is described by these models. Therefore, there is some understanding, explanation or mechanism at the lower levels (*e.g.* Cell division). They possess the capability to mimic suitable physical, chemical or biological processes and answers how and why a particular response occurs. The modeler usually begins with some empirism and with the attainment of knowledge additional parameters and variables are commenced to elucidate crop yield. The system is therefore broken down into components and assigned processes.

3. Static and Dynamic Models

The static model does not comprise time as a variable although the end products of cropping systems are assembled over time. On the contrary dynamic models absolutely incorporates time as a variable and maximum dynamic models are primarily expressed as differential equations.

4. Deterministic Models

A deterministic model makes exact predictions for quantities such as crop yield or rainfall, without any corresponding probability distribution, variance or random element. However, variations due to inaccuracies in recorded data and to heterogeneity in the material being dealt with are inherent to biological and agricultural systems (Brockington, 1979). In some cases, deterministic models may be appropriate in spite of these inherent variations but in others they might prove to be inadequate *e.g.* in rainfall prediction. The deterministic models become more inadequate with greater uncertainties in the system.

5. Stochastic Models

When variation and uncertainty extends at a high level, it is suggested to establish a stochastic model that offers an expected mean value as well as the associated variance. However, these models incline to be technically inconvenient to handle and can quickly become complex. Hence, it is beneficial to attempt to resolve the problem with a deterministic approach initially and to attempt the stochastic approach only if the results are inadequate and unsatisfactory.

6. Simulation Models

These models are designed with the motive of imitating the behaviour of a system. The aspect of variability associated with diurnal change in weather and soil conditions is amalgamated as these models are designed to resemble the system at short time intervals (daily time-step). Enormous amount of input data such as climate parameters, soil characteristics and crop parameters should be accessible to run the model. They provide the probability of specifying management options and can be utilized to explore an extensive range of management strategies at low costs.

7. Optimizing Models

These models possess a particular objective of evolving the ideal option concerning management inputs for practical operation of the system. To derive solutions, they adopt decision rules that are compatible with some optimizing algorithm. This enforces a bit of rigidity into their structure which results in restricting the representation of stochastic and dynamic aspects of horticultural systems.

Applications of Crop Models

Application of simulation modeling is increasing in teaching, research, farm and resource management, policy analysis and production forecasts. It can be applied in crop system management, research and policy analysis.

1. *Research understanding*: Model development assures the amalgamation of research understanding attained through cautious disciplinary research and enables the recognition of the crucial factors that operates the system. It also highlights the areas where knowledge is inadequate. Therefore, more targeted and efficient research planning can be done by adopting a modeling approach.
2. *Amalgamation of knowledge across disciplines*: Basic research carried out in different regions, countries and continents can be integrated through adoption of modular framework. This ensures a minimization of research costs (*e.g.*, through a diminution in duplication of research) furthermore collaborating researchers at an international level.
3. *Refinement in experiment documentation and data organization*: Development, testing and application of simulation model demands utilization of a huge amount of technical and observational data provided in given units and in a specific order. Data handling forces the modeler to resort to formal data organization and database systems.
4. *Site-specific experimentation*: Crop models can be implemented to foresee crop performance in regions where the crop has not been grown before or not grown under optimal conditions.
5. *Yield analysis*: Adoption of a model with a sound physiological background makes it possible to extrapolate to other environments. Simulation models are employed to climatically-determine yield in various crops. Quantification of yield reductions caused by non-climatic causes such as delayed sowing, crop spacing, soil fertility, pests and diseases, also becomes possible through the modeling approach. Models have also been recorded as useful tool in partitioning yield gains into components caused by genetic improvements, changing weather trends and improved technology.
6. *Climate change projections*: Climatic variability especially the aberrant weather conditions is one of the major concerns of scientific as well as general community. The implementation of crop models to analyse the impact of climate change has been widely used across the globe. The temperature of the earth is increasing due to increased concentration of carbon dioxide and other greenhouse gases. Crop production immensely

depends on alternation in weather and therefore any change in global climate have major effects on crop yields and productivity. Elevated temperature and carbon dioxide affects the biological processes like respiration, photosynthesis, plant growth, reproduction, water use *etc.* Proper understanding of the effects of climate change will therefore help scientists to guide farmers to make crop management decisions such as selection of crops, cultivars, sowing dates and irrigation scheduling to minimize the risks.

7. *Scoping best management practices*: Simulation can be performed to establish the best management practices under a definite cropping system. In the past, the chief focus of research was on crop production. Recently, along with profitable crop production, the environment quality has become an essential issue that horticultural producers must address. Producers require strategies for maximizing the profitability of crop production along with maintaining soil quality and reducing environmental degradation. Solutions to this new challenge demands consideration about interaction of numerous components that effects plant growth. To accomplish this objective, future research will involve considerably more effort and resources than present research activity. Models possessing chemical leaching or erosion components can be employed to establish the best long-term farming practices. Investment decisions like purchase of irrigation systems can be taken with an eye on long term usage of the equipment when irrigation schedules are done using the modeling approach.
8. *Yield forecasting*: Reasonably precise estimates of crop yield over large areas prior to actual harvest are of immense importance to both the researcher and the farmer in terms of planning. In this approach the model is run using actual weather data during the cropping season for the geological region of interest. Weather years for typical years are used to continue simulations until harvest.

Model Parameterization: Calibration, Evaluation and Validation

Model calibration requires the refinement of a few model parameters such that data simulated by the error free model fit the observed data. In several instances, despite a model is based on observed data, simulated values do not exactly comply with the observed data and slight modifications have to be made for some parameters. Sampling errors and incomplete knowledge of the system may give rise to noncompliance. Alternatively, it may arise when the model is applied in a situation that is remarkably different from the one under which it was developed.

The model validation phase involves the verification that the calibrated model closely represents the actual situation. This procedure comprises of a comparison between simulated output and observed data that have not been previously used in the calibration stage. due to lack of detailed datasets it is not possible to validate all the components, therefore, validation of only the determinant ones are adopted. For example, in a soil water crop model, it becomes essential to validate the extractable water and leaf area components since biomass accumulated is thoroughly dependent on these factors. Evapotranspiration also becomes a determinant to validate.

Crop Models: Constraints

Crop models are unable to give correct projections due to insufficient understanding of natural processes and computer power limitation. As a result, the evaluation of possible effects of climate change are predicted on estimations. Moreover, most models are not able to provide reliable projections of changes in climate variability on local scale, or in frequency of exceptional events such as storms and droughts (Shewmake, 2008). General Circulatory Models (GCMs) so far have not been able to create reliable projections of alternations in climate variability, such as shift in drought and storms frequencies, although these could considerably affect crop yields.

Misuse of models may occur as different users possess varying degrees of expertise this field. As crop models are not universal, the user has to select the most suitable model according to his objectives. GCMs do a reasonable job in simulating global values of surface air temperature and precipitation, but do poorly at the regional scale (Grotch, 1988).Moreover, biological and agricultural models are reflections of systems for which the behaviour of some components is not fully understood and differences between model output and real systems cannot be fully accounted for. Methodology of model validation is still rudimentary.

The major reason is that, unlike the case of disciplinary/traditional experiments, a huge set of hypothesis is being assessed concurrently in a model. The affirmation of models nowadays is further complex by the fact that field data are rarely so precise that validation can be conclusive. This results from the fact that model parameters and driving variables are obtained from site-specific situations that preferably should be measurable and available. However, practically, soil, plant and meteorological data are scarcely accurate and may come from nearby locations. Sometimes, parameters that were not routinely measured may emerge to be essential and they are then arbitrarily approximated. Inherent soil heterogeneity over relatively small distances and to variations arising from the effects of husbandry practices on soil properties leads to variation in measured parameters. Crop data reflects soil heterogeneity as well as variation in environmental factors over the growing period.

Performance of a model is restricted to the quality of input data. It is common in cropping systems to possess large volumes of data concerning to the above-ground crop growth and development, but data concerning to root growth and soil characteristics are usually not as extensive. Most simulation models require that meteorological data be absolute and authentic. Eventually, sampling errors also adds to inaccuracies in the observed data.

An ultimate crop model would be one that physically and physiologically defines all relations between variables the model reproduces and universally real world behaviour. However, such a model cannot be developed because the biological system is too complex and many processes involved in the system are not fully understood (Jame and Cutforth, 1996). Although, an absolute crop model could be created, the assembling of the extremely accurate system parameters and of the input data for the crop environment would be an intimidating task in itself. Thus, the extent of detail involved in a crop model is closely associated to the end use of the model and the precision required. Even when a judicious choice is made, it is essential that aspects of model limitations be borne in mind such that modeling studies are put in the proper perspective and successful applications are achieved.

REFERENCES

Brockington NR.1979. Computer Modeling in Agriculture. Clarendon Press New York, 156 pp.

de Wit CT. 1965. Photosynthesis of leaf canopies. Agricultural Research Report No. 663 PUDOC, Wageningen, The Netherlands.

de Wit CT (1967). Photosynthesis: its relationship to overpopulation. p.315-320. In Harvesting the sun. Academic Press, New York.

de Wit CT and Goudriaan J. 1978. Simulation of assimilation, respiration and transpiration of crops. Simulation monograph. PUDOC, Wageningen, The Netherlands.

Goudriaan J. 1977. Crop micrometeorology: a simulation study. Simulation Monograph. PUDOC.Wageningen, The Netherlands.

Grotch SL.1988. Regional Intercomparison of General Circulatory Model Predictions and Historical Climatic Data. U.S. Dept of Energy, Office of Energy Research Report, DOE/NBB-0084 (TR041) pp 291

Jame YW. 1992. Modeling soil moisture for agricultural water management. Pages 69–88. In F. J. Eley, R. Granger and L. Martineds. Soil moisture modeling. Environment Canada, Ottawa ON.

Jame YW and Cutforth HW 1996. Crop Growth Models for Decision Support systems. Canadian J. Plant Sci. 76(1): 9-19. http: //pubs.aic.ca/doi/abs/10.4141/cjps96-003. Assessed 1st August, 2012

Jones JW 1993. Decision support systems for agricultural development. Pages 459–471 in F.W.T. Penning de Vries, P. Teng and K. Metsellaar, eds. Systems approaches for agricultural development. Kluwer Academic Publishers, Dordrecht, The Netherlands.

Kumar R and Chaturevdi S. 2009. Crop Modeling: A Tool for Agricultural Research. Agropedia. http: //agropedia.iitk.ac.in/?q=content/cropmodel. Accessed 1st August, 2012

Linnemann H, de Hoogh J, Keyzer MA, van Heemst HDJ. 1979. MOIRA, Model of International Relations in Agriculture. Report of the Project Group, Food for a Doubling World Population, Amsterdam.

Penning de Vries FWT, Brunsting AB and van Laar HH 1974. Products, requirements and efficiency of biological synthesis, a quantitative approach. *J. Theor. Biol.* 45: 339-377.

Shewmake S. 2008. Vulnerability and the Impact of Climate Change in South Africa's Limpopo River Basin. IFPRI Discussion paper 00804. http: //books.google.com.gh/books?id Accessed 1st August 2012

Tsuji GY, Uehara G and Balas S. 1994. DSSAT: a decision support system for agrotechnology transfer. Version 3. Vols. 1, 2 and 3. University of Hawaii, Honolulu, HI.

Chapter 7

Coffee Cultivation in the Non-traditional Bastar Region of Chhattisgarh

Krishan Pal Singh[1], Beena Singh[2] and D.S. Thakur[3]

[1,2]Scientist (Horticulture) and [3]Dean,
College of Horticulture and Research Station, IGKV, Raipur – 494 001, Chhattisgarh
[2]Scientist (Horticulture), S.G. College of Agriculture and Research Station,
IGKV, Raipur – 494 001, Chhattisgarh

Today coffee is grown commercially in the continents of South America, Asia and Africa. The top coffee producing countries in world is Brazil, Columbia, Indonesia, India and Ethiopia. Coffee is one of the most developing and world's largest exports after oil as its revenues are vital to many nations. As an agro-based rural enterprise primarily, this industry is a source of employment for over a million people in cultivation, processing and trade sectors. Coffee cultivation is also instrumental in preserving the precious forest ecosystem in traditional areas while in the non- traditional areas it was introduced to check *podu* or shifting cultivation whereby to control denudation of forest and also soil erosion.

The beauty of Bastar often termed as the tribal district of Chhattisgarh, lies in its natural forest area and various types of tribal population. Before splitting into seven districts, Bastar was one of the largest districts in India, with an area of 39,114 sq km, which was even greater than the Kerala state and some other countries like Belgium, Israel *etc.* The district is abundantly and richly endowed with forest resources that comprise more than 70 per cent of the total land area. Bastar zone lies at an altitude of 500-1276m above mean sea level and the soil ranges from shallow to deep and friable with a pH from 4.5 to 7.0. The forest soils are very rich in nutrients. The rainfall ranges from 1300-1800 mm with frequent showers during summer season. Darbha block of Bastar is located at 18°52′ N and81°55′E at an elevation of 678 m above MSL. The soil and climatic conditions of Bastar is quite similar to the southern states of the country where coffee is cultivated in large areas. Such a climatic condition is well suited for higher growth and yield of the crop.

Coffee is a shade loving plant. It is mainly cultivated in deep fertile jungle soils under a two tier mixed shade canopy comprising of evergreen leguminous and non-leguminous shade plants. Shade plants are beneficial in various aspects as they provide a natural habitat for various flora and fauna that contributes towards the soil fertility and reduces soil erosion. The horticultural practices followed in the Indian coffee plantations are considered as one of the best in the world, in which emphasis is mainly towards manipulation of microclimate and plant health, so as to reduce excessive dependence on agrochemical inputs.

Coffee production has been commercialized for the first time in Darbha block of Bastar zone in Chhattisgarh. Intercropping of mango, silver oak and jackfruit has also been done to provide additional shade to the crop. Presently four varieties of Arabica and one variety of Robusta have been planted in the orchard. Seeds of these varieties have been procured from the Coffee Board, Koraput. The work has been undertaken in the mixed forest land with an aim of improving the livelihood of the tribal population. The economic prospects and commercial benefits of the crop are attracting more number of youth and rural farmers towards itself.

Tid Bits of Coffee

Soil and Climate

The coffee soils in India belong to the red and lateritic soil groups. They differ in texture from sandy loam to clayey loam. The soil pH of 6.5 is best suited to the crop. Whenever, the coffee soil pH falls below 6.1, the soil pH should be corrected by the application of alkali forming soil amendments like agricultural lime (calcium carbonate) and dolomite (Calcium and magnesium carbonate). Highly acidic soils are harmful for the coffee plant growth as the plant gets stunted in nature. If the soils are alkaline most of the essential plant nutrients are not available to the plant. Liming of the coffee soils can be done anytime of the year except during monsoon period. For good growth of coffee, the soil should be deep (75 cm), friable, well drained, rich in organic matter and slightly acidic.

Table 7.1: Soil and Climatic Conditions Required for Coffee

Factors	*Arabica*	*Robusta*
Elevation	1000-1500 m	500-1000 m
Soils	Deep, friable, rich in organic matter, well drained and slightly acidic	Deep, friable, rich in organic matter, well drained and slightly acidic
Slopes	Gentle to moderate	Gentle slope to fairly level
Temperature	15-30°C	20-38°C
RH	70-80 per cent	80-90 per cent
Annual rainfall	1600-2500 mm	1000-2000 mm
Blossom showers	March- April (20-35 mm)	Feb-March (20-35 mm)
Backing showers	April-May (50-75 mm)	March-April (40-60 mm)

Source: Singh *et al.*, 2019.

Coffee is a tropical plant which is also grown in semi-tropical climate. A temperature range of 20-27°C is suitable for the crop with an annual rainfall of 100-200 cm. Growth is rapid during hot rainy season and cool dry season. Such a climate is suitable for berry ripening.

Table 7.2: Major Climatic Factors Affecting Production in Coffee

Factors	*Arabica*	*Robusta*
Blossom showers	Absence of rain till May	Absence of rain till March
Backing showers(if delayed)	Poor crop growth	Poor crop set
Hail stones	Plant injury	Plant injury
Rain at blossom	Partial to full failure	Partial failure
Soil moisture	Partial to full flowering failure	Complete flowering failure
Water logging	Partial to complete crop failure	Partial to complete crop failure
Winds	Highly affected	Partial affected

Source: Singh *et al.*, 2019.

Varieties of Coffee

a. *Coffea arabica*

The plant produces profuse branches and the matured leaves are dark green in color while the young leaves are either green or bronze. The flower buds are mainly produced in clusters in the axils of leaves at each node. Water is an important factor for flowering. Blossoms usually occur 8 to 10 days after showers. It is considered to be superior in quality over *Coffea robusta* for having rich flavour and aroma. It is self fertile and is susceptible to major pests and diseases *viz.*, coffee leaf rust and white stem borer.

b. *Coffea robusta*

The plants of Robusta are green in colour. Its flowers are fragrant and borne in compact clusters at each node. The number of flowers per node is comparatively higher than Arabica. The flower normally opens on the 7th or 8th day after showers. Robusta is self-incompatible. The ovule of the same flower does not get fertilized with its own pollen hence cross pollination is essential in this variety. It takes nearly 10 to 11 months for fruit maturity. Robusta generally get ready for harvest two months later than Arabica. These are shallow rooted crops and are thus susceptible to drought but are tolerant to major diseases and pests like coffee leaf rust and white stem borer. The quality of Robusta coffee is inferior to that of Arabica in bean size as well as in organoleptic characteristics like aroma, flavour and taste.

c. Plant Coffee (*Coffea liberica* Bull ex Hiern)

Liberica is known as 'Plant coffee' because of its growth habit. It is cultivated along the fence and boundary line of estates. The leaves are dark green, leathery and are 2 to 3 times larger than Arabica. The flowers are large and fleshy. The plant usually takes about a year to start flowering. The fruits are larger in size than Arabica and yellow to reddish-brown in colour.

Propagation

Propagation in coffee is mainly done by seed, cutting and grafting. Although propagating through seed is the simplest and easiest to adopt by the growers. Seed sowing should be done in the month of December - January. In coffee two types of shoots *viz.*, horizontal (plageotropic) and vertical (orthotropic) shoots are produced. Orthotropic shoots are suitable for vegetative propagation. By adopting vegetative propagation method, uniform plants with desirable characters like higher yield, superior quality as well as resistance to pests and diseases could be ascertained in Robusta and also in hybrids of Arabica.

Types of Planting Design

1. Square system
2. Hedge row planting
3. Paired row planting

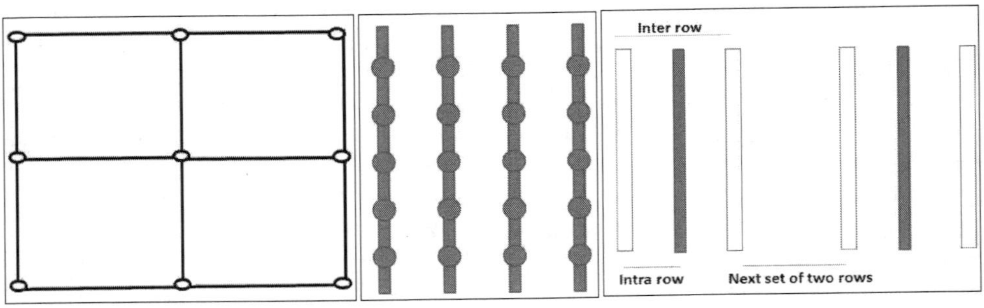

Square System **Hedge Row Planting** **Paired Row Planting**

Planting of Shade Plant

Coffee is a shade loving plant. It gives a superior quality and taste when grown under shade. Shade plants are therefore recommended as a protective measure in areas of harsh environmental conditions such as high temperature, extended drought, heavy rainfall, high winds and occurrence of hail storm. Shade trees can be planted as a shade belt in East- West direction to protect the coffee plants from southern exposure at a spacing of 6 m apart within a row and 12 m between the two rows. Permanent shade plants are planted at wider spacing (9 to 12 m) wherever the forest plant cover is inadequate. Under unirrigated conditions the Arabica coffee requires 50 to 60 per cent filtered shade and Robusta coffee requires upto 40 to 50 per cent shade while under irrigated condition nearly 30 to 40 per cent shade gives consistent production.

Nutrient Management in Coffee

Studies have indicated that the peak period of demand for nutrient are the time of flowering, fruit set and development and maturation of the crop. Hence fertilizer application is suggested during blossom (February–March), pre monsoon (May–June), monsoon (August), and the post monsoon period (September–October) in areas under Southwest monsoon. Every metric ton of clean Arabica coffee produced needs roughly about 15 to 20 kg of sulphur while, Robusta needs 20 to 25 kg.

Table 7.3: Recommended Dose of Fertilizers in Coffee

Year	Pre-blossom March $N:P_2O_5:K_2O$	Post-blossom May $N:P_2O_5:K_2O$	Mid-monsoon August $N:P_2O_5:K_2O$	Post-monsoon October $N: P_2O_5 K_2O$	Total
1st year	15:10:15	15:10:15	-	15:10:15	45:30:45
2rd and 3rd year	20:15:20	20:15:20	-	20:15:20	60:45:60
4th year	30:20:30		-	30:20:30	80:60:80
5 years	40:30:40	40:30:40	20:0:0	40:30:40	140:90:120

Source: https://www.indiacoffee.org.

Table 7.4: Required Amount of Manure in Coffee

Age of Plants	*Amount*
2nd Year	0.7 kg/plant
3rd Year	1 kg/plant
4th Year	2 kg/plant
5th Year onwards	kg/plant

Weed Control

In young orchards, weeds pose a serious problem particularly during the initial three years. Free growth of weeds reduces the yield of coffee and plant resumes a sickly appearance. If weeding is not done during October-November then it should be positively undertak en during December using slash weeding. Use of herbicides should be done judiciously in order to maintain the soil and environmental health from hazardous chemicals. In the case of chemical weeding, use either contact weedicide like Gramoxone or systemic weedicides like Glycel or Round Up in a rotation.

Irrigation Management

Coffee is primarily cultivated as a rainfed crop worldwide. It requires nearly 5 to 10 liters water per plant once in a week after the monsoon period. This avoids the wilting and the death of young plants and facilitates easy and faster rooting of the plants. Drip and sprinkler irrigation are well suited for the coffee orchards. Robusta is quite susceptible to drought therefore; irrigation must be done after one month of monsoon.

Harvesting the Cherries

The fruits also known as coffee cherry turn to bright red colour when ripe and are ready to be harvested. It takes around 3 to 4 years for the newly planted coffee plants to bear fruits. It can be harvested in two ways:

1. **Strip picked:** All of the cherries are stripped off from the branch at a time, either by machine or by hand.
2. **Selectively picked:** Only the ripe cherries are harvested, and they are picked individually by hand. Pickers rotate among the plants every 8 to 10 days choosing only the cherries which are the peak of ripeness. Because this kind of harvest it is labour intensive and more costly. It is used primarily to harvest the finer Arabica beans.

Processing of Beans

Once the coffee has been picked, processing must begin as quickly as possible to prevent fruit spoilage. Depending on the location and local resources, coffee is processed in one of the two ways:

1. Wet method of processing (Parchment coffee preparation) - Parchment coffee or plantation coffee is prepared by pulping cleanly washed beans with water thereby drying under sun. The quality of parchment coffee is always superior to that of cherry coffee (whole fruit dried under sun). Cherry coffees always give fruity taste as the beans are in contact with the mucilage for a longer time during the course of drying period.
2. Dry method of processing (Cherry coffee preparation) - To prepare cherry coffee, fruits should be picked as and when they are ripe. Greens and under ripe fruits should be sorted out and dried separately. All the extraneous matters like twigs and leaves should be separated from the fruits and dried. The fruits should be spread uniformly in the drying yards at a thickness of 7 to 8 cm clean cement or clay lined drying yards should be used for drying the cherry coffee. Coffee should be stirred and ridged at least once every hour to facilitate uniform drying. In the evening time Cherry coffee should be heaped and covered with polythene or tarpaulin sheets to avoid contact with the mist. The coffee will be fully dry in about 10 to 12 days under the bright weather conditions. Cherry coffee should be dried to a moisture level of 12 to 12.5 per cent.

Diseases and Pests in Coffee

Arabica and Robusta coffee are susceptible to fungal diseases. Bacterial and viral diseases have not been recorded so far in India. Basically Arabica coffee is more susceptible to diseases than Robusta.

Major Diseases of Coffee

- ✰ Leaf rust – *Hemileia vastatrix* B. and Br.
- ✰ Black rot – *Kolero ganoxia* Donk.

- ☆ Anthracnose – *Colletotrichum gloeosporioides*Penz.
- ☆ Coffee root diseases
    - i. Brown root disease – *Fomes noxious* Corner.
    - ii. Red root disease –*Poria hypolateritia* Berk.
- ☆ Coffee trunk canker – *Ceratocystis fimbriata* Ell. and Halst.
- ☆ Berry blotch – *Cercospora coffeicola* B. and Cke.
- ☆ Nursery diseases
    - i. Collar rot – *Rhizoctonia solani* Kuhn.
    - ii. Brown eye spot –*Cercospora coffeicola* B. and Cke.
    - iii. Stem necrosis and leaf spot – *Myrothesium roridum* Tode ex Fries

Major Insect-Pests of Coffee

- ☆ Coffee white stem borer- *Xylotrechus quadripes* Chevrolat
- ☆ Coffee berry borer – *Hypothenemus hampei* Ferrari
- ☆ Shot hole borer or Black twig borer – *Xylosandrus compactus* Eichhoff
- ☆ Mealy bugs – *Planococcus citri*
- ☆ Green scale – *Coccus viridis*
- ☆ Cockchafers or White grub – *Holotrichia* spp.
- ☆ Brown scale – *Saissetia coffeae* (Walker)
- ☆ Hairy caterpillar- *Eupterote* spp.
- ☆ Root lesion nematode – *Pratylenchus coffeae*
- ☆ Thrips – *Heliothrips haemorrhoidalis* (Bouche)
- ☆ Termites – *Odontotermes* spp.

REFERENCE

Singh, K.P., Singh Beena, Kerketta, A. K. and Sahu, S (2019) Coffee in Chattisgarh, IGKV/Technical Bulletin/2019-20.

Chapter 8

Livelihood Generation and Skill Development in the Tribal Dominated Bastar Belt through Coffee Cultivation

Beena Singh[1] and K.P. Singh[2]

[1]Scientist (Horticulture), S.G. College of Agriculture and Research Station, IGKV, Raipur – 494 001, Chhattisgarh

[2]Scientist (Horticulture), College of Horticulture and Research Station, IGKV, Raipur – 494 001, Chhattisgarh

The State Chhattisgarh lies within 17°45′ to 20°34′ N latitude and 80°15′ to 82°15′ E longitude with an altitude ranging between 550 and 850 meters above the mean sea level (MSL). It covers a geographical area of 1,37,360 sq km of which nearly 65.90 per cent of the total area is covered by tribals and hence often recognized as tribal dominated State. The State has a good scope of irrigation, expansion, use of ground water and rainwater conservation. The 'Bastar Plateau Zone' of Chhattisgarh includes the districts *viz.*, Bastar (Jagdalpur), Bijapur, Dantewada, Kanker, Kondagaon, Narayanpur and Sukma. This zone includes the Dandakaranya region, plains of Mahanadi, Indravati and Sabri rivers tract, which are separated by the hills of Abujhmar, Bailadila, Tikampalli, Usoor and Gola-palli and is surrounded by the states of Maharashtra, Andhra Pradesh and Odisha on three sides. Except Indravati river plains, most of the area is covered by evergreen dense reserve forests and hilly tracts. The major landforms are high-level plateaus, structural hills and valleys and pediments and pediplains. The sub agro-ecological region of Bastar Plateau includes the North Lowland, Keshkal Escarpment, Abujhmad Hills, North-East Plateau, Southern Plateau and Southern Lowland.

Bastar district lies in its natural forest area with various types of tribal community. The Bastar District is abundantly and richly endowed with forest resources (45-60 per cent). The forests in this district can be divided in to four belts, namely, Northern mixed forests, Central moist region comprising of Sal belts, teak belt zone and the dry region comprising mixed forests. The major types of forest in Bastar division are Sal forests, teak forests and mixed forests. The major portion of the forest area is covered with Sal forest which mostly lies on the plateau. In the hilly region, Sal is confined to narrow valleys and on lower slopes.

Podu cultivation by felling of forest has been the general practice of the tribal population in the projected area. Agricultural crops like Ragi, Sama, Paddy, Rajmash, Niger and other pulses are also raised in the area for one or two seasons. After cultivation of the above crops for one or two seasons the farmers move on to the nearest new forest areas. This leads to the depletion and destruction of the forest area. The practice of podu/shifting cultivation has caused denudation of forests, which is a serious threat to the ecology, resulting in soil erosion and siltation of the river basins.

The substantiation for encouraging coffee plantations in the lands of tribal is based primarily on the livelihood and economic sustainability. It is therefore useful to analyze the relative returns which are likely to accrue to the farmers on taking up various crops. Through coffee plantation the farmers are capable of generating returns which are higher than the returns from other agricultural crops. Coffee performs better when planted as an intercrop rather than a monocrop. The crop is mainly cultivated in deep fertile jungle soils under a two tier mixed shade canopy comprising of evergreen leguminous and non-leguminous shade plants. When growing under shade it has several advantages. Shade plants provide a natural habitat for vast population of birds and natural enemies of insect pests/diseases that helps in reducing the soil erosion, contribute towards the fertility of coffee soils by recycling nutrients from deep soil in the form of leaf litter and finally protect the coffee bushes from vagaries of changing weather conditions.

In India, coffee has mainly been cultivated in larger areas in the southern states. The soil and climatic condition of Bastar region is quite similar to the southern states of country. Keeping this in view coffee cultivation was targeted in Darbha block of Bastar District with a view of generating employment as well as livelihood to the tribal population. This is the foremost attempt that has been made in the barren forest lands to commercialize the production of coffee in Chhattisgarh. The soil of Bastar ranges from shallow to deep, friable with a pH range from 4.5 to 7.0. Altitude varies from 500 to 1276 meters above mean sea level while the rainfall is quite high which ranges from 1300-1800 mm with frequent showers during summer season. The forest soils are very rich in nutrients. Darbha is situated at 18°52′ N 81°55′E and at an elevation of 678 m above the mean sea level. The coffee orchard is a mixed forest area where plantation of jackfruit, mango and silver oak has also been done to provide additional shade along with the cultivation of crops like black pepper, guava, oilseeds and leafy vegetables that can provide an extra income to the grower. Economic upliftment of the farmers was kept as priority while initiating the project. This in turn would also increase the standard of living of the rural youth and help lowering the migration of localites in search of job.

The cost-effectiveness outlook and profitability of coffee is attracting people towards its production in the zone. This work is also a measure to increase the forest cover and to provide ecological and environmental balance in the tribal forest tracts. The work posed to be bit challenging being a non traditional area for the crop plantation and LWE affected district. The initial approach of convincing the rural youth is the major factor that determines the possibilities of raising any crop in the farmer's field. Most of the rural population being illiterate and staying in remote areas are not interested in changing their cropping pattern easily. But due to the constant efforts of the researches and scientists working in the area, recently few farmers have taken up the cultivation of coffee as a part of their farming system. By frequent trainings they are being taught about the modernization in farming that is now-a-days convincing them to adapt the technologies that could help them improve their standard of living by catching up the current agriculture practices. The coffee plantations in the lands of tribal has been found to provide sustainable income to the tribal farmers and the cultivation of coffee is seen to give relatively better income to the farmers than other crops. As a result of their adaptiveness the cultivation is now being taken up at 100 acres in the Dilmili block of Bastar too.

With the advent of this piece of work marketing linkages of processed coffee among SHG's and National agencies would be established which in turn would provide employment and livelihood security to the tribal farmers. A set of four varieties of *Coffea arabica viz.*, Sam Ramon, S-8, S-9 and Chandragiri dwarf alongwith *C. robusta* variety CxR have been planted and evaluated for yield and its attributes. Utilizing further left out interspaces/canopy spaces in coffee plantation areas by various intercrops *viz.* black pepper pulses, oilseeds, leafy vegetable *etc.* is known to amplify the year round production of crops as well as help combating the initial income of the farmers. Along with these benefits introducing processing products like filter coffee, green coffee as well as branding of the same as well as other intercrop produces would provide extra income generation. As of now there is no coffee processing unit in the Bastar zone till date. This will alternatively give a way for the establishment of coffee processing unit in the zone as well thereby boosting employment too.

Bringing considerable improvements through the project:

1. Through coffee cultivation the farmers of the area could fetch an income of Rs. 15000 – 20000/- per acre (after 4th year of crop) in the non-productive barren lands. Further an additional income of about Rs. 15000 per acre (after 4th year of crop) could be generated through various fruit and spices intercrops and Rs. 10000 per acre (from 1st year) from various other intercrops in initial stages of intervention which may enhance during later years. Thus a total of Rs. 45000 per acre net income can easily be generated from lands which are left unproductive throughout the year.
2. Deforestation issues could be checked.
3. Popularizing non-traditional crops like coffee and black pepper is fetching high level interest among tribal farmers of nearby areas too.

4. Year round availability of sufficient skilled manpower for labour intensive operations like manual weeding, shade regulation and soil conservation measures and the like. Thereby generating year round employment.
5. The horticultural practices followed in the Indian coffee plantations are considered as one of the best in the world, in which emphasis is mainly towards manipulation of microclimate and plant health, so as to reduce excessive dependence on agrochemical inputs.

Impacts Resulting from the Project

1. Additional annual revenue to the State/Central Government through taxes.
2. Afforestation and maintenance of ecological balance.
3. Alteration of waste lands into productive lands.
4. Creation of industries for financial support and market centers which would generate additional employment opportunities, resulting into automatic development of infrastructure.
5. Effective soil conservation, improving soil fertility as well as reduction in soil erosion.
6. Employment generation and sustainable livelihood.
7. Permanent settlement of tribal families and improvement of their socio-economic standards so as to ensure social stability and economic security.
8. Reducing migration of tribal farmers to the adjoining States in search of livelihood.
9. Skill development in the rural youth as well as the farming community.

Chapter 9

Salinity Stress and Salicylic Acid: Its Consequences and Regulation in Horticultural Crops

***Radha Singh*[1]*, Reena Nair*[2] *and J. Jyothsna*[3]**

[1]Scientist, Plant Physiology, College of Agriculture, Rewa, JNKVV
[2]Assistant Professor and [3]Ph.D. Scholar
Department of Horticulture, College of Agriculture, JNKVV, Jabalpur

Soil salinity is a serious agricultural problem worldwide because it limits plant growth and productivity. High salinity stress is the most severe environmental stress, impairing crop production. High salinity interferes with plant growth and development and can also lead to physiological drought conditions and ion toxicity (Zhu, 2002). Therefore, high salinity affect mostly all aspects of plant physiology and metabolism and cause both hyperosmotic and hyperionic stresses, which lead to plant demise. High salt deposition in soil leads to a deposition of a low water potential zone in the soil. This makes it increasingly difficult for the plant to acquire water as well as nutrients. Salinity causes ion-specific stresses resulting in an altered K^+/Na^+ ratio. The external Na^+ can negatively impact intracellular K^+ influx and salinity leads to a build up of Na^+ and Cl^- concentrations in the cytosol, which can be ultimately detrimental to the cell. The Na^+ can dissipate the membrane potential and therefore facilitates the uptake of Cl^- down the gradient. Higher concentrations of sodium ions (above 100 mM) are toxic to cell metabolism and can inhibit the activity of many essential enzymes, cell division and expansion, membrane disorganization, and osmotic imbalance, which finally can lead to growth inhibition. Higher concentrations of sodium ions can also lead to a reduction in photosynthesis and the production of reactive oxygen species. Potassium ions are one of the essential elements required for growth. Alterations in K^+ ions (because of the impact of high salinity stress) can disturb the osmotic balance, the function of stomata, and the function of some enzymes. High salinity can also injure cells in transpiring leaves, which leads to growth inhibition. This salt-specific or ion-excess effect of salinity causes a toxic effect of salt inside the plant. The salt can concentrate in the old leaves and the leaves die, which is crucial for the survival of a plant (Munns *et al.*, 2006). Growth reduction is linked with the time over which the plant has grown in saline conditions. Over all soil salinity is a major problem in agricultural areas all over the world that limits plant growth and productivity (Yildirim *et al.*, 2009; Qin *et al.*, 2010).

Salt stress can lead to stomatal closure, which reduces CO_2 availability in the leaves and inhibits carbon fixation, exposing chloroplasts to excessive excitation energy which in turn increase the generation of reactive oxygen species (ROS) such as superoxide, hydrogen peroxide, hydroxyl radical and singlet oxygen (Parida and Das, 2005; Ahmad and Sharma, 2008; Ahmad *et al.*, 2010a, 2011). In many plant studies, it was observed that production of ROS is increased under saline conditions (Hasegawa *et al.*, 2000) and ROS-mediated membrane damage has been demonstrated to be a major cause of the cellular toxicity by salinity in different crop plants such as rice, tomato, citrus, pea and mustard (Gueta-Dahan *et al.*, 1997; Dionisio-Sese and Tobita, 1998; Mittova *et al.*, 2004; Ahmad *et al.*, 2009, 2010b). Many morpho-physiological, biochemical and molecular studies have revealed the deleterious effect of salinity stress on plants. Therefore, emphasis must be given on different crops for their sustenance under stress by evolving techniques for improving the physiological and biochemical status of plants.

SA application is done by several ways *viz.*, soaking the seeds prior to sowing, adding to the hydroponic solution, irrigating, and spraying solution, to protect various plant species against abiotic stress factors by inducing a wide range of processes involved in stress tolerance mechanisms (Horvath *et al.*, 2007).

The role of SA in defence mechanism to alleviate salt stress in plants was studied by Afzal *et al.*, 2006 and Hussein *et al.*, 2007. The ameliorative effects of SA have been well documented including salt tolerance in many crops such as bean (Azooz, 2009), tomato (Tari *et al.*, 2002). Root drenching with 0.1 mM SA protected tomato (Lycopersicon esculentum) plants against 200 mM NaCl stress (Stevens *et al.*, 2006). It was found that the SA treatment caused accumulation of both ABA and IAA in wheat seedlings under salinity. However, the SA treatment did not influence on cytokinin content. Thus, protective SA action includes the development of anti-stress programmes and acceleration of normalization of growth processes after removal of stress factors (Sakhabutdinova *et al.*, 2003). The results obtained in the last few years strongly establish that SA could be a very promising conpound for the reduction of abiotic stress sensitivity of crops, because under certain conditions, it has been found to mitigate the damaging effects of various stress factors in plants.

The effect of salicylic acid (SA) on morpho-physiological, biochemical parameters, changes in antioxidant enzymes activity as well as yield and yield contributing parameters in several horticultural crops is reviewed as under:

A. Morpho-Physiological Parameters under Salinity Stress as Influenced by Salicylic Acid

1. Germination

Seed germination is one of the most elementary and vital phases in the growth cycle of plants that find out plant establishment and the yield of crops. Salinity have many-fold effects on the germination process: it alters the imbibitions of water by seeds due to lower osmotic potential of germination media causes toxicity which alters protein metabolism (Yupsanis *et al.*, 1994; Dantas *et al.*, 2007), disturbs hormonal balance (Khan and Rizvi, 1994), and reduces the utilization of seed reserves (Promila and Kumar 2000; Othman *et al.*, 2006). It may also negatively affect the ultrastructure of cell, tissue and organs (Koyro, 2002; Rasheed, 2009). The reduction in germination percentages due to increasing soil salinity may be attributed to membrane damage and stimulate formation of ROS such as superoxide, hydrogen, hydroxyl radical (Wong *et al.*, 2009). However, there are various internal (plant) and external (environmental) factors that affect seed germination under saline conditions which includes nature of seed coat, seed dormancy, seed age, seed polymorphism, seedling vigour, temperature, light, water and gasses (Wahid *et al.*, 2011).

The available literature revealed the effects of salinity on the seed germination of various crops like Pea (Shahid *et al.*, 2012; Ahmad *et al.*, 2017), Brassica spp. (Ibrar *et al.*, 2003; Ulfat *et al.*, 2007) and hot pepper (Khan *et al.*, 2009). It is well established that salt stress has negative correlation with seed germination and vigour (Rehman *et al.*, 2000). Similar reduction pattern in germination percentage with the rise in NaCl concentrations have been observed in bean, pea (Hamada and Enany, 1994). Higher level of salt stress inhibits the germination of seeds while lower level of salinity induces a state of dormancy (Khan and Weber, 2008).

Seeds of halophytes and nonhalophytes respond in a similar way to salinity stress, *e.g.*, the initial germination process are delayed under salt stress (Almansouri *et al.*, 2001; Khajeh-Hosseini *et al.*, 2003). The inhibitory effect of NaCl stress on seed germination is due to an osmotic effect and/or ion toxicity (Rehman *et al.*, 1996; Katembe *et al.*, 1998; Pujol *et al.*, 2000; Tobe *et al.*, 2004), and salinity-induced declines in germination are usually due to an osmotic effect only for halophytes, but nonhalophytes are more likely to exhibit additional ion toxicity (Bajji *et al.*, 2002). Akhtar and Hussain (2009) also observed a reduction in the germination percentage of Vicia faba seeds due to increasing soil salinity from 5 to 15 dS m^{-1}. Overall, salinity through enhancement of osmotic pressure leads reduction of water absorbance and disturbance in metabolic and physiological processes will be under its effect.

Several reports have been published in the last decade demonstrating the role of salicylic acid (SA) in seed germination. The SA-induced increase in the germination rate observed in the presence of NaCl may also be due to the activation of the second stage of germination, which corresponds to the embryo growth stage, through enhanced synthesis of proteins that are essential for germination (Rajjou *et al.*, 2006) and the activation of the mobilization of reserve metabolites with low molecular weights (Nonogaki *et al.*, 2010). LEAs and HSPs are synthesized in the presence of exogenous SA and have been shown to enable germinating Arabidopsis seeds to show a marked tolerance to salt stress (Rajjou *et al.*, 2006). SA significantly stimulated the activities of enzymes involved in germination such as transkelolase, enolase, malate dehydrogenase, phosphoglycerate kinase, glyceraldehyde 3-phosphate, dehydrogenase, fructose 1,6-diphosphatase, and pyruvate decarboxylase (Eastmond and Graham, 2001; Rajjou *et al.*, 2006).

Khan and others (2009) studied the effect of seed priming with salicylic acid and acetyl salicylic acid on the seed germination and seedling growth of hot pepper cv. Hot Queen. While the salinity affected the germination and seedling growth, the acetyl salicylic acid at 0.2mM as seed priming treatment for 48 hours, exhibited improvements on Final emergence Percentage (FEP), Emergence index (EI), shoot length, root length and vigour index (VI). The authors also highlighted that towards the salinity tolerance induction, the acetyl salicylic acid is a more promising candidate than the salicylic acid itself.

Ahmad *et al.* (2017) carried out a study involving priming of pea seeds with various concentrations of salicylic acid (0.5 mM and 1.00 mM) and NaCl (50, 100 and 150 mM). The salinity did reduce the germination of non-primed seeds whereas the seeds treated with 1.00 mM of SA and 150 mM of NaCl showed the maximum improvement in the germination percentage (24.6 per cent). Similarly, Ahmad and others (2020) reported 33 per cent hike in the germination percentage when the sweet pepper seeds were primed with 0.2 mM of SA and 60 mM NaCl.

2. Growth Parameters

Soil salinity is a major peril in agriculture sector that affects the major morpho-physiological processes in plants at different growth stages. High salinity influences plants in several ways: water stress, ion toxicity, nutritional disorders, oxidative stress, alteration of metabolic processes, membrane disorganization, reduction of cell division and expansion, genotoxicity (Hasegawa *et al.*, 2000; Munns, 2002; Zhu, 2007). The high sodium levels disturb potassium (K) nutrition and when accumulated in cytoplasm that inhibits many enzymes (Sharma and Manchanda, 1997). High concentrations of salts in the root zone decrease soil water potential and the availability of water (Lloyd *et al.*, 1989). Murphy *et al.* (2003) reported that salt stress affects plant physiology at both whole plant and cellular levels through osmotic and ionic stress. The role of SA in increasing the relative water content (RWC) and thereby increasing the water use efficiency (WUE) of the plants were reported in Yildirim *et al.* (2008) and Dong *et al.* (2011) in cucumber, Mirdad *et al.* (2014) in broccoli, Hemantaranjan *et al.* (2014) and Ahmad *et al.* (2017) and Ahmad *et al.* (2020) in pea and Husen *et al.* (2018) in mustard.

The treatment with SA ameliorated the plant physiological traits such as chlorophyll fluorescence, stomatal conductance, net photosynthetic rate, transpiration rate and intracellular CO2 concentration in mustard, pea, strawberry and Dianthus superba (Husen *et al.*, 2018; Ahmad *et al.*, 2020; Feghih *et al.*, 2017; Ma *et al.*, 2017). With the foliar spray of 1.00 mM SA, Rady and others (2015) recorded improved leaf anatomical characteristics namely increased mid-vein length, midvein width, median vascular bundle width, median vascular bundle length, blade thickness, palisade tissue thickness and spongy tissue thickness in common beans.

An increase in the leaf area plant-1, so the leaf area index was observed in Calendula (Bayat *et al.*, 2012), common beans (Rady *et al.*, 2014), broccoli (Mirdad *et al.*, 2014), eggplant (Raghami *et al.*, 2016), pea (Hemantaranjan *et al.*, 2014; Singh *et al.*, 2015), strawberry (Mohamed *et al.*, 2017), tomato (Souri *et al.*, 2019) and sweet pepper (Ahmad *et al.*, 2020)

Zhu *et al.* (2001) reported that salinity stress causes reduction in internodal distance and number of leaves plant^{-1} in various plant species and this reduction may be attributed due to the reduction in turgor potential, necessary for cell elongation (Iqbal and Ashraf, 2005). Production of ABA associated with salinity stress that acts as senescing agent could be thus, a major cause for reduced number of leaves under salinity stress. Albacete *et al.* (2008), Ghanem *et al.* (2008) and Shahid *et al.* (2012), investigated that Solanum lycopersicum, Solanum tuberosum and pea respectively under saline conditions and found a discernible reduction in number of leaves and internodal distance in response to salinity stress. Shareef (2019) reported that the exogenous SA application of SA increased the endogenous levels of auxins (IAA), cytokinin and gibberellins (GA_3) and this established the role of SA in regulating the plant hormones.

There has been always an endeavour of researchers as well as of farmers as to how to combat the situation of salinity stress, which on the one hand being expanded rapidly in several agricultural segments and on the other hand causing multifaceted damage to crop plants with regard to plant growth and development. By and large, salinity stress mitigation could be possible either by taking controlling measures for reducing soil salinity or through strengthening susceptible and tolerant plant species by treating seeds or seedlings by innovative techniques of pre-soaking/ hardening or foliar application of select chemicals/growth regulators. Further, cultivation of select tolerant varieties could be the ideal way, but up to a limited degree of salinity stress. This hypothesis might be carried out through screening of a number of plant growth regulators effective at different degrees of salinity stress under different agro-climatic conditions coupled with several tolerant, semi-tolerant and moderately tolerant varieties. Amongst the applications of plant growth regulators in the recent past, certain characteristics as influenced by salicylic acid under abiotic stresses are encouraging.

Salicylic acid is a phenolic molecule act as a plant growth regulator modulating plant responses to environmental stresses. Salicylic acid has been reported to play role in mitigating the harmful effects on morpho-physiological parameters of salinity stress. Induction of multiple stress tolerance in plants by exogenous application of salicylic acid and its derivatives may have a significant practical application in agriculture, horticulture and forestry (Senaratna *et al.*, 2000). Salicylic acid treatment alleviated the deleterious effect of salinity, drought and boron toxicity on plant growth. The effects of salicylic acid either alone or in combination with salt stress has been reviewed as given below:

Lots of findings indicate that salicylic acid responses are concentration dependent. The lower concentrations of salicylic acid enhanced the growth parameters such as root and shoot length, leaf number and leaf area *etc.*, which were reported in eggplant (Raghami *et al.*, 2016), hot pepper(Khan *et al.*, 2009) and okra (Esan and Olaiya, 2016).

Besides the significance of SA concentrations, effects of salicylic acid on salt tolerance of maize could be determined by measuring the growth parameters: shoot and root lengths, shoot and root fresh and dry weights and leaf area. It was found that salicylic acid enhanced the salt tolerance in terms of improving the measured plant growth and other growth parameters.

Bayuelo *et al.* (2008) found that the salinity stress reduced the growth of *Phaseolus vulgaris*. Shahid (2011) reported that application of NaCl @ 25, 50 and 75 mM in pea genotypes, causes significant reduction in embryo axis length (radical + plumule), root and shoot dry weight. Under the SA treatment, the growth parameters *viz.*, the fresh and dry weights of shoot and root were ameliorated in many crops, under salinity stress. The investigation carried out by Yildirim and others (2008) concluded that the fresh weight of shoot and root; dry weight of the shoot and root of the cucumber plants were higher than the plants that didn't receive the SA spray. Singh and others (2015) found that the fresh weight and dry weight of roots and shoots were higher in the plants treated with SA 0.25 mM as soil application. Raghami and others (2016) investigated that when eggplants were exposed to two different levels of NaCl, *i.e.*, 75 and 100 mM. Results indicated that the plant height, number of leaves, leaf area, fresh and dry weight of shoots, and fresh and dry weight of roots were drastically affected by the salinity imposedSimilarly, effects promoting higher fresh and dry weight of the plant parts through seed priming were reported in pea (Hemantaranjan *et al.*, 2014) and sweet pepper (Ahmad *et al.*, 2020), This explains that exogenously applied salicylic acid either with seed soaked (SS) or soil incorporated (SI) or foliar spray (FS) increased plant growth significantly under a number of abiotic stress induced situations. An increase in the biomass was recorded with radish plants when the plants under salt stress were exposed to foliar spray of 0.2 mM SA. It is also exciting to note down that exogenous SA treatments ameliorated the negative effects of salt stress on fresh and dry weights of certain useful plants as reported by Stevens *et al.* (2006) and Szepesi *et al.* (2005) in tomato and Karlidag *et al.* (2009) in strawberry. Nevertheless, an increase in growth parameters of salt affected plants in response to SA might be related to the protective role of SA on membranes that could increase the tolerance of plants to salt stress (Wang *et al.*, 2006; Aftab *et al.*, 2010a).

Bayat and other (2012) conducted an experiment imposing foliar spray of 1.00 mM SA along with 200 mM NaCl as soil drench to the Calendula plants. The reports proved that the growth traits namely the plant height and leaf area plant-1 were increased by the SA treatment. In the same way, foliar spray of 3.00 mM SA increased number of crowns plant, number of leaves plant and leaf area plant in strawberry (Mohamad *et al.*, 2017). Further, Syeed *et al.* (2011) studied the influence of salicylic acid (SA) on two mustard (*Brassica juncea* L.) cultivars, *viz.*, Alankar (salt tolerant) and PBM16 (salt sensitive), differing in the degree of salt tolerance, *i.e.*, foliar application of SA @ 0.1, 0.5, or 1.0 mM was made at 15 DAS on NaCl-grown plants (NaCl @ 0 and 50 mM) and concluded that 0.5 mM SA increased maximum leaf area and dry mass in both cultivars compared to control, however maximum leaf area was noted in Alankar than PBM16. Similar reports on SA's ability to alleviate salinity stress were studied by Ahmad *et al.* (2017), Mohamed *et al.* (2017), Husen *et al.* (2018) and Shahmoradi *et al.* (2018) in pea, strawberry cv. Festival, mustard and jasmine, respectively.

The effects of acetyl salicylic acid were also reported in sweet pepper (Khan *et al.*, 2009) and tomato (Sajyan *et al.*, 2019). Khan and others (2009) conducted the experiment that involves various treatments of NaCl stress (3 and 6 dSm^{-1}) and 0.8 mM salicylic acid and 0.2 mM acetyl salicylic acid as seed priming of hot pepper cv. Hot Queen. The results stated that the acetyl salicylic acid outperformed the salicylic acid, by imparting higher levels of salinity tolerance through improved shoot and root length under salinity stress of 6 dSm^{-1}. Similarly, increased leaf number $plant^{-1}$, plant height and stem diameter were witnessed in tomato, under 100 ppm SA as foliar spray and salinity stress of 8 dSm^{-1}.

3. Membrane Stability Index

Salinity impaired membrane stability by increasing electrolyte leakage. The integrity and functions of biological membranes are very sensitive to environmental stress and stress-induced damage to membranes has been well documented in the plants (Nishida and Murata, 1996). It has been noted that maintaining integrity of cellular membranes under stress conditions is considered an integral part of salinity tolerance mechanisms. The membrane stability index (MSI) could be considered as a reliable physiological parameter to screen germplasm lines for salinity tolerance (Turner and Nicolas, 1987).The higher leakage of solutes was probably due to enhanced H_2O_2 accumulation and lipid peroxidation under salt stress (Dionisio-Sese and Tobita 1998). The presence of NaCl in the rooting medium caused a disturbance in membrane permeability expressed by an increase in solute leakage in sugar beet and the leakage was higher in susceptible cultivars of sugar beet than tolerant ones, indicating severe membrane damage for the former under salt stress (Ghoulam *et al.*, 2002). The plants fed with NaCl exhibited a significant increase in electrolyte leakage compared to the control. Similar results were reported by Kaya *et al.* (2003), Karlidag *et al.* (2009) and Khan *et al.* (2010). Shahid *et al.* (2012) reported that, Membrane stability index (MSI) decreased under salt stress in all the pea genotypes at all NaCl treatments (0, 25, 50, 75 mM) but maximum reduction was noted under 75 mM NaCl. Membranes damage increased with increase in salinity level, so MSI can be considered as very significant tool for evaluating the salt tolerance potential in pea genotypes. However, application of salicylic acid (SA) partly maintained membrane stability. Salicylic acid (SA) reduced electrolyte leakage in cucumber hypocotyls under chilling stress (Kang and Saltveit, 2001). Similarly SA reduced the amount of ion leakage (measured as electrolytes) in salt stressed tomato plants indicating that SA treatment has facilitated the maintenance of membrane functions (*i.e.*, semipermeability) under stress conditions (Stevens *et al.*, 2006). The periwinkle plants supplemented with NaCl (50, 100, 150 mM) exhibited a significant increase in electrolyte leakage. However, foliar spray of salicylic acid (10-5 M) reversed the adverse effect of NaCl and caused a significant decrease in electrolyte leakage in the salt-treated plants (Idrees *et al.*, 2011). The sweet pepper plants were subjected to seed priming of 0.2 mM and NaCl stress of 20, 40, and 60mM. As an effect of

SA, the plants exhibited 21.3 per cent reduction in the electrolyte leakage than the control plants (Ahmad *et al.*, 2020). Similar results reported in lemongrass (Idrees *et al.*, 2012), cucumber (Yildirim *et al.*, 2008), Gerbera (Kumara *et al.*, 2010), Faba bean (Orabi *et al.*, 2013) and Calendula (Bayat *et al.*, 2012).

Hemantaranjan and others (2014) primed the seeds of pea with 1.5 mM of SA and exposed to salinity stress. Under 150 mM of salinity, the primed seedling showed increased membrane stability index than the untreated seedlings. The reports of Chaparzedah *et al.* (2015) were in line with the above result as the membrane stability of radish plants were maintained by 0.2 mM SA at 180 mM salinity.

B. Biochemical Parameters under Salinity Stress as Influenced by Salicylic Acid

1. Photosynthetic Pigments

Salinity stress causes decrease in plant growth and productivity by disrupting physiological processes, primarily photosynthesis. The accumulation of intracellular sodium ions at salt stress changes the ratio of K: Na, which affect the bioenergetic processes of photosynthesis (Sudhir and Murthy, 2004). A number of findings suggested that reduction in chlorophyll content under salinity stress causes reduction in photosynthesis and consequently low yields. Fernanda *et al.* (2004) found that salt stress negatively affected leaf stomatal conductance and transpiration rate. At ultra structural level, only thylakoid swelling and a decrease in the amount of grana stacking were observed in treated plants. The higher concentration of NaCl causes significant reduction in total chlorophyll in pea (Ahmad and Jho, 2005). Shahid *et al.* (2011) used ten different genotypes of pea were grown in pots having fine sand as growth medium. After 30 days of germination, the plants were subjected to salt stress under 0, 25, 50 and 75 mM NaCl. After one week of salt application, photosynthesis rate, stomatal conductance, transpiration rate, and chlorophyll contents of the plant were remarkably decreased with increasing salinity in all the genotypes.

Salicylic acid (SA) ameliorated the adverse effect of salinity stress on photosynthetic pigments. Chlorophyll content increase with SA treatment confirmed the reports of Yildirim *et al.* (2008) in cucumber, Bayat *et al.* (2012) in Calendula, Hadi *et al.* (2014) in common bean, Sunaina *et al.* (2014) in cabbage, Hemantaranjan *et al.* (2014) in pea, Singh *et al.* (2015) in pea and Ma *et al.* (2017) in Dianthus. An increase in the Chlorophyll a, b and total chlorophyll contents were observed when the radish plants were sprayed with 0.2 mM SA (Chaparzedah *et al.*, 2015). Five seeds of tomato were planted in pot and kept in a growth chamber under controlled conditions. Then the pots were treated with 5 levels of NaCl (0, 25, 50, 75, 100 mM) and 4 levels of salicylic acid (0, 0.5, 1 and 1.5 mM). The obtained result showed that 50mM salinity concentration significantly increased total chlorophyll and 1.5mM SA concentration led to increase in total chlorophyll (Zahra *et al.*, 2010).

Besides chlorophyll, the SA application has also reported to produce increase in accessory pigments such as carotenoids and anthocyanin. In a hydroponic culture of cabbage, wherein the NaCl concentration was maintained at 150 mM, foliar application of 1.00 mM SA resulted in increased carotenoid content in the cabbage leaves. Similar trend is noticed in eggplant (Raghami *et al.*, 2016), mustard (Husen *et al.*, 2018), pea (Singh *et al.*, 2015), radish (Chaparzedah *et al.*, 2015) and Dianthus (Ma *et al.*, 2017). Similarly, SA application also increased the anthocyanin content in leaves of radish (Chaparzedah *et al.*, 2015) and fruits of strawberry cv. Pajaro (Jamali *et al.*, 2013). The increase in the accessory pigments relates to the strengthening of the antioxidant abilities of the plant to overcome the salinity stress.

2. Carbohydrates Metabolism

Soluble sugars participated in the osmotic adjustment of plants exposed to the salt stress. It is assumed that during germination and early seedling growth, cell division and enlargement require transportation of respiratory substrates in the form of soluble sugars and low molecular weight protein from seed storage organs to the site of growth (Bewley and Black, 1994). Salinity stress limited hydrolysis of food reserves from storage tissues as well as it impaired their translocation from storage tissue to developing embryo axes (Dubey, 1985; Lacerda *et al.*, 2003). As reported, tolerant genotypes accumulated more soluble carbohydrate than did sensitive ones. Both ionic and non-ionic stresses increased the concentration of reducing sugars and sucrose. Tolerant varieties accumulated sucrose at a significantly greater level than did sensitive ones under non-ionic stress condition (Ildiko and Gabor, 2000). The stressed plant showed increase in soluble sugars as the concentration of salt was increased as reported in pea (Ahmad *et al.*, 2005). Dkhil and Denden (2010) reported that NaCl suppressed significantly starch content in both cotyledons and embryonic axes. It has been reported that salt stress limit the mobilization of starchy endosperm reserves in several species, as a result of inhibition of different enzymatic activities (Lin and Kao, 1995; Almasouri *et al.*, 1999).

Certainly, SA causes balance in the sugar level at salinity stress condition due to possibility of increase in induced glucose storage by salt stress. That is for storage demand reduction of carbon or starch decomposition (Tattini, 1996). Increase of all soluble carbohydrate in the root during salinity stress is effective on the balance against osmotic pressure. The plant cell for escaping from plasmolysis performance and creation during salt stress conditions should be changed and analysed from macro molecule to micro molecule Sucrose breaks down to glucose and fructose, and starch decomposition to glucose increases its osmotic pressure cell (El Midaoui *et al.*, 1999). Maria (2000) and colleagues reported increase in soluble sugar content of the root of tomato under salt (NaCl) stress. The use of salicylic acid could activate the consumption of soluble sugar metabolism by increasing osmotic pressure. It is supposed that salicylic acid treatment deranges the enzymatic system of polysaccharide hydrolysis (Khodary, 2004). Szepesi (2006) stated that in SA applied plants leaves accumulated more compatible osmolytes, *e.g.*, soluble sugars, glucose and fructose, a sugar alcohol, sorbitol and proline. The exogenous application of SA counteracted the drought deleterious effects on sugars in maize (Khodary, 2004) and okra (Amin *et al.*, 2009). Salvia officinalis (garden sage, common sage) subjected to NaCl concentration of 0,4,8 and12 dS m^{-1} and salicylic acid concentration of 0,1,2, and 4 mM. Results showed higher concentrations of leaf sugar at highest salt concentration, which were induced further by salicylic acid @ 4 mM. The highest rise in sugar level was recorded when salinity was 12 dS m-1 along with salicylic acid concentration. The least sugar level was observed at 0 mM salinity (0 dS m-1) 0 mM of salicylic acid (Sahar *et al.*, 2011). Subsequently, Azooz *et al.* (2011) observed in broad bean seedlings that foliar application of salicylic acid (@ 0.0, 0.5 and 1.0 mM) enhanced the soluble sugar content in roots and shoots and 25 per cent seawater treatments supplemented with the 1.0 mM SA to the plants have recorded highest levels of soluble sugars.

An investigation made by Dong and others (2011) involved a hydroponic system conditioned with salinity stress and various doses of SA. They reported that after SA treatment, the levels of glucose, fructose, raffinose and stachyose were increased in both leaves and roots. In contrast, sucrose and starch responded differently in leaves and root, decreasing in leaves but accumulating in roots. These changes could be due to the effects of SA on the activities of metabolism-related enzymes. In leaves, SA promoted the activities of sucrose phosphate synthase (SPS), sucrose synthase (SS) and amylases, while the SA-treated root showed a reduced amylase activity and an unchanged activity in SPS. The changes in various sugar contents resulted in the accumulation of soluble sugars in SA-treated cucumber seedlings, especially non-reducing sugars in roots. In Shara (Plectranthus tenuiflorus) plant SA enhanced the soluble sugar accumulation in both water-stressed and non-stressed plants (Jalal *et al.*, 2012). The starch content decreased under saline conditions.

Esan and Olaiya (2016) investigated the effect of SA in Okra, at various concentrations *viz.*, 10-2, 10-4, and 10-6 mM along with NaCl conditioning of 50, 100 and 150 mM. They recorded a significant increase in the accumulation of reducing sugars (glucose and fructose) upon pre-soaking with 10-4 mM SA and 50 mM NaCl solution. In like manner, pre-soaking with 1.00 mM SA and 50, 75 and 100 mM NaCl solution resulted in increase of sugar content in pea by by 58.1 per cent, 130.2 per cent and 312.2 per cent, respectively (Ahmad *et al.*, 2017). Similar reports on increase in total soluble sugar content by SA were made by Hadi *et al.* (2014) and Rady *et al.* (2015) in common bean, Sunaina *et al.* (2014) in cabbage, Raghami *et al.* (2016) in eggplant, Shareef (2019) in date palm and Ahmad *et al.* (2020) in pea.

3. Protein Metabolism

One of the mechanisms affected by salinity stress in plants is protein synthesis. It is known that soluble protein content is an important indicator of physiological status of plants. The alternation of protein synthesis or degradation is one of the fundamental metabolic processes that may influence water stress tolerance (Ouvrard *et al.*, 1996; Jiang and Huang, 2002). Both quantitative and qualitative changes of proteins have been detected during the stress (Riccardi *et al.*, 1998; Ahire *et al.*, 2005; Kottapalli *et al.*, 2009). Diverse findings indicate that reduction in protein content has been observed under salinity stress and this has been attributed to reduced synthesis and accelerated proteolysis. It may be due to decreased incorporation of amino acids into the proteins and denaturation of amino acids and protein synthesis under saline conditions (Levitt, 1972; Bernstein, 1974). Total soluble protein content decreased in plants under NaCl stress reported in Lycopersicon esculentum, Vicia faba, Amaranthus tricolor and Brugiera parviflora (Alaghabary *et al.*, 2004;; Gadallah, 1999; Parida and Das, 2005; Parvaiz and Satyavati, 2008; Wang and Nil,

2000). Doganlar *et al.* (2010) studied the effect of salt stress on pigment and total soluble protein content in different varieties of tomato (Lycopersicon esculentum Mill.) who found that after 48 h of salt application total soluble protein content of all tomato cultivars were significantly decreased.

Amini *et al.* (2007) reported that salt treated tomato plants showed one of the up-regulated proteins that was designated as a salt tolerant protein with a mol. wt. of 25 kDa. It has already been reported that salt and osmotic stresses can increase the expression of stress proteins. Plants produce proteins in response to abiotic and biotic stress and many of these proteins are induced by Phytohormones including salicylic acid (Naz, 2008). Azooz (2009) reported that in faba bean genotypes salicylic acid treatment increased the total protein in both root and shoot. Sahar (2011), said that salicylic acid increases protein concentration and the highest level of protein content observed at 4 mm SA level with 12 dS/m salinity level in Salvia officinalis. Application of salicylic acid increased total protein in Rauwolfia serpentina Leaves @ 100 mM NaCl (Misra and Misra, 2012).

Sunaina and other (2014) carried out an experiment with cabbage under hydroponic culture. They found that the protein content of the leaves and head increased with increasing concentration of SA namely 0.05 and 0.10 mM under salinity stress. The presoaking treatment of pea seeds in 1 mM SA significantly improved the protein content by 12.59 per cent, 48.1 per cent and 66.9 per cent with 50, 100, 150 mM NaCl salinity respectively (Ahmad *et al.*, 2017). Similar reports on increased protein content by SA were made in common beans, jasmine and tomato (Hadi *et al.*, 2014; Shahmoradi *et al.*, 2018; Souri *et al.*, 2019). Besides, SA application was also found to lower the protein degradation in strawberry cv. Selva (Jamali *et al.*, 2016).

Some plant proteins are synthesized during the stress and are responsible for stress alleviation. The glycine betaine and proline betaine are the salinity induced stress-proteins whose mechanism of stress alleviation takes place through increasing the cell solute potential, thereby increasing the uptake of water by the roots. Esan and Olaiya (2016) studied the effect of SA on stress-protein synthesis under salt stress. The priming of okra seeds with 10-4 mM SA favoured the accumulation of osmoprotectants namely glycine betaine and proline betaine in the leaves. Jamali and others (2016) also reported an increase in glycine betaine synthesis and its accumulation in the leaves of strawberry cv. Selva.

4. Nitrogen Metabolism

Nitrate reductase is the key enzyme in nitrogen metabolism, known to be regulated by various environmental factors apart from its own substrate, nitrate. Many research showed the effect of salinity stress on nitrate reductase activity in different plants. Gouia *et al.* (1994) while working on bean observed that when the two plants were subjected to salt stress, biomass was greatly reduced in beans. In addition to growth, nitrate reductase activity proved much sensitive to salinity in case of beans as well as cotton. Flores *et al.* (2000) reported that salinity inhibits nitrate uptake and nitrate reductase activity in tomato seedlings. Debouba *et al.* (2006) found that when tomato plants (*Lycopersicon esculentum* Mill, cv. Chibli F1) grown for 10 days on control medium were exposed to different concentrations of NaCl (0, 25, 50, and 100 mM). Increasing salinity led to a significantly decreased NR activity in leaves at 50 and 100 mM NaCl. Bybordi *et al.* (2011), studied the effect of different concentrations of salinity (@ 0, 50, 100, 150 and 200 mM NaCl) on nitrate reductase activity and other enzymes activity along with nitrate and phosphate uptake and total nitrogen content in leaves of canola (*Brassica napus* L.) and the team reported that nitrate reductase activity was increased significantly in plants subjected to 100 mM NaCl, whereas it was decreased gradually under 150 and 200 mM NaCl treatment as compared to control treatments.

Excessive amount of sodium ions in plant cells causes enzyme inhibition of nitrogen metabolism (Soussi *et al.*, 1998), *i.e.*, nitrate reductase (NR) (Shafea, 2003) as well as photosynthetic enzymes rubisco and PEP carboxylase (Soussi *et al.*, 1999) followed by metabolic dysfunction (Booth and Beardall, 1991) such as degradation of photosynthetic pigments (Abdullah and Ahmed, 1990; Soussi *et al.*, 1999). Inhibition of nitrate reductase (NR) activity is caused due to salinity stress or other environmental stresses were also reported by many workers (Khan *et al.*, 1990; Ashraf *et al.*, 1995). The process of nitrate reduction is well known for its sensitivity to environmental stresses (Iqbal *et al.*, 2006). Nitrate reductase enzyme play a key role in nitrogen metabolism and NR is associated with protein synthesis and plant growth. In Brassica juncea NR activity increased by the dilute concentration (10-5 M) of SA (Fariduddin *et al.*, 2003). Pandey and Agarwal (2002) suggested that NR activity can rapidly change in response to several environmental factors, including low temperature and osmotic stress. Dar *et al.* (2007) reported in their experiment that NR activity decreased in leaves due to induced salinity (@ 50,100 and 200 mM NaCl) and the reduction in activity was proportional to the concentration of NaCl in the growing medium. The maximum reduction in activity was noted respectively at 20 and 40 days after sowing at 200 mM NaCl as compared to control. However, hardening of seeds by SA (@ 1.0 mM) increased NR activity and the interaction between 200 mM NaCl and 1.0mM SA reduced the loss in activity of NR. Jakab *et al.* (2007) proposed that SA increased the availability of nutrients by enhancing turgor of plants under stress.

NR is an inducible enzyme and its concentration depends on the availability of substrate so improvement in NR activity in foliar applied SA was due to the availability of more substrates (Akhtar *et al.*, 2013). Under a hydroponic culture of cabbage, wherein different concentrations of SA and salt stress were imposed, the foliar application of 1.5 mM SA exhibited increased nitrate reductase activity at 50 mM of NaCl concentration in growing medium while the untreated and plants under lower SA doses showed reduced nitrate reductase activity. Husen and others (2018) investigated the effects of SA in mustard, at various concentration of NaCl. They reported that the SA application at 0.5 mM as foliar spray increased the synthesis and accumulation of nitrate reductase under 200 mM NaCl conditioning. Besides, the SA application maintained the nitrate reductase activity under high salinity stress in pea (Hemantaranjan *et al.*, 2014).

5. Proline Metabolism

Proline plays an important role in mediating osmotic adjustment and protecting the sub-cellular structures in stressed plants. Arora and Saradhi (1995) reported that the seedlings exposed to stress accumulated higher levels of proline in their shoots compared with control seedlings. Proline and quaternary ammonium compounds are key osmolytes that help plants to maintain cell turgor (Hsu *et al.*, 2003; Seki *et al.*, 2007). A large number of plant species accumulate proline in response to salinity stress and their accumulation may play a role in defense against salinity stress. However, data do not always indicate a positive correlation between osmolytes accumulation and an ability to adapt to stress (Ashraf and Harris, 2004; Mansour, 2000; Mansour *et al.*, 2005). Proline may act as a signaling component of the adaptation process (Maggio *et al.*, 2002). The effect of long-term (30 days) NaCl treatments (100 mM and 200 mM) on the accumulation of proline was studied in potato (Solanum tuberosum cv.) leaves. Salt stress positively affected proline accumulation in treated plants (Abraham *et al.*, 2003; By *et al.*, 2004). Agarwal and Pandey (2004) reported that proline concentration in many salt tolerant plants has been found to be higher than in salt sensitive plants. A lot of findings suggested that proline accumulation rapidly increased in plants in response to salinity stress as found in case of Pisum sativum (Ahmad and Jhon, 2005; Ahmad *et al.*, 2008), Phaseolus aureus (Misra and Gupta, 2005) and Morus alba (Ahmad *et al.*, 2007). Shahid *et al.* (2011) investigated the performance of two salt tolerant (climax and samarinazard) and two salt sensitive (ambassidar and PF-400) pea genotypes under salt stress. Seeds of all tested genotypes were exposed to four different salinity levels, *i.e.*, control, 25, 50 and 75 mM NaCl. Hoagland solution was used as nutrient medium throughout the study. Salt stress significantly reduced the leaf proline in all pea genotypes as compared to control. The salt tolerant genotypes contained maximum leaf proline than sensitive genotypes. Similar findings was reported by Haid *et al.* (2014), Rady *et al.* (2014) in common beans, Sunanina *et al.* (2014) in cabbage, Chaparzedah *et al.* (2015) in radish, Souri *et al.* (2019), Jamali *et al.* (2016) in strawberry and Ma *et al.* (2017) in *Dianthus superba*. AI-Saady *et al.* (2012) investigated that leaf proline in fenugreek (*Trigonella foenum-gracium* L.) plants, which were treated with different levels of salt (@ 0, 4, 6, 8, 10 dSm-1) increased significantly over the control plants and maximum leaf proline was measured at 8 dSm-1 in fenugreek.

Proline is one of the most important osmoprotectant of plants defensive mixed action to salt stress. Increasing proline leads to increase in resistance to salt stress, but the amount of this increase is different between different varieties (Sairam *et al.*, 1998). Proline can be considered as a ROS scavenger, and plays a protective role in cellular structures, which can break down and provide carbon, nitrogen and energy sources after abiotic stress exposure to plants. Source of proline is total protein and total amino acid. Several findings are available about the increase of proline in plants exposed to salinity stress (Ehsanpour and Fatahian, 2003; Parida and Das, 2005; Hussein *et al.*, 2007; Koca *et al.*, 2007; Eraslan *et al.*, 2007; Sakhanokho and Kelley, 2009). According to the other studies, the proline accumulation by salicylic acid treatment increases in wheat under oxidative stresses, (Tasgin *et al.*, 2006). The more tolerant plants store more proline (Desnigh and Kanagaraj, 2007). Sahar *et al.* (2011) conducted an experiment with Salvia officinalis seeds planted in pots containing perlite were put in a growth chamber under controlled conditions of 27 ±2°C and 23 ±2°C temperature, 14h light and 10h darkness; NaCl concentration of 0,4,8,12 ds m-1 and salicylic acid concentration of 0,1,2,4 mM were used in the form of factorial experiment in a complete randomized design (CRD) and observed the salicylic acid effect on the sugar, protein and proline contents under salinity (NaCl) stress showed that with increasing salinity levels, proline increase, which also increases further by high salicylic acid level. This elucidates that there is a direct relation between accumulation of osmolytes (proline and sugars) and increasing resistance in plants under abiotic stress condition (Ramanjulum *et al.*, 1998). Remarkably, high protection of abscisic acid in plants treated with salicylic acid under salt stress increases proline and defensive proteins (Shakirova, Sakhabutdinova, 2003). Amin *et al.* (2009) reported that increasing the amount of proline and sugars in plants would lead to the resistance against loosing water, protect turgor, reduce the membrane damage and accelerate the growth of plants under stress conditions. The higher accumulation of proline under stress conditions was attributed to enhanced activities of proline biosynthesis enzymes, ornithine aminotransferase and pyrroline-5-carboxylate reductase, as well as due to inhibition of proline degradation enzymes, proline oxidase and proline dehydrogenase (Kishor *et al.*, 2005). Shahba *et al.* (2010) explained that salt stress may also induce osmotic stress as well as sodium toxicity and accumulation of osmoprotectants such as proline. Rezayatmand *et al.* (2012) found that when SA (@ 0, 0.5 and 1.5 mg L^{-1}) applied exogenously in the presence of NaCl (@ 0, 50 and 150 mM) on Artemisia aucheri in vitro culture, salt stress increased proline along with soluble and reducing carbohydrates levels.

The pre-soaking of pea seeds with 1 mM SA and 150 mM NaCl resulted in increased accumulation of proline under salinity stress (Ahmad *et al.*, 2017). Similarly, the foliar spray of 0.2 mM SA improved proline content of radish leaves under salt stress (Chaparzedah *et al.*, 2015). Upon root treatment with SA, the tomato plants also showed increased proline while the untreated plants recorded with lower proline contents (Souri *et al.*, 2019).

C. Antioxidant Enzymes Activity under Salinity Stress and Influenced by Salicylic Acid (SA)

Induction of antioxidant enzyme activities is a general adaptation strategy which plants use to overcome to the deleterious effects of salinity stress. It is well known that salinity induced oxidative stress in plants (Ashraf, 2009; Noreen *et al.*, 2009). Salt-induced oxidative stress in plants generate different types of enzymatic and non-enzymatic antioxidants. Commonly known enzymatic antioxidants include superoxide dismutase (SOD), ascorbate peroxidase (APX), catalase (CAT), peroxidase (POD) and glutathione reductase (GR), whereas non-enzymatic antioxidants are glutathione (GSH), ascorbate (AsA), carotenoids and tocopherols.

Environmental stresses induce production of reactive oxygen species (ROS), *viz.*, superoxide radicals, hydrogen peroxide (H_2O_2) and hydroxyl radicals (OH^-). These ROS cause oxidative damage to different cellular components including membrane lipids, protein and nucleic acids (Halliwell and Gutteridge, 1986). Reduction of oxidative damage could provide enhanced plant resistance to salt stress. Plants use antioxidants such as reduced glutathione (GSH) and different enzymes such as superoxide dismutases (SOD), CAT, APX, glutathione-S-transferases (GST) and glutathione peroxidases (GPX) to scavenge ROS. Plants have the antioxidant enzymes for overcoming the harmful effects of salinity stress such as SOD, CAT, APX *etc.* Gomez *et al.* (2004) found increase in all SOD isoenzymes of pea chloroplasts following a long term NaCl treatment. Ahmad *et al.* (2008), carried out experiment in two varieties of pea and reported in pea (*Pisum sativum* L.) plants, when plants were supplemented with 50, 100, 150 and 200 mM NaCl, SOD and APX activity significantly increased in both the varieties over the control. The first line of defense in oxidative stress is the action of SOD that converts O_2–• to H_2O_2 (Mittler, 2002; Ashraf, 2009).

The increase in H_2O_2 is very harmful for the cells; hence it gets dismutated to water and oxygen with the help of another antioxidant enzyme catalase (Breusegem *et al.*, 2001; Ashraf, 2009). One of the adaptive traits under salinity stress is increase in catalase activity which reduces the toxic levels of H_2O_2 and protects the cell from oxidative damage (Sekmen *et al.*, 2007; Vital *et al.*, 2008; Ashraf, 2009; Noreen *et al.*, 2010). Another versatile antioxidant enzyme is ascorbate peroxidase which utilizes ascorbate (AsA) as electron donor and scavenges H_2O_2 in water-water and ascorbate glutathione cycles. Hydrogen peroxide is reduced to water by APX and plays a role in cell defense mechanism (Kangasjarvi *et al.*, 2008; Ashraf, 2009). Type of differential response of cultivars differing in salt tolerance has already been observed in a number of crops, in which salt tolerant cultivars exhibited higher antioxidant enzyme activities than did the salt sensitive cultivars (Ashraf, 2009; Noreen *et al.*, 2009; Sabir *et al.*, 2011). Noreen and Ashraf, (2009) reported that in six different pea cultivars under different saline regimes, CAT enzyme activity increased in salt tolerant pea, which was found to be a reliable marker of salt tolerance. Ahmad *et al.* (2012), found that SOD, CAT, APX activity induced in three cultivars of mustard (*Brassica juncea* L.) under salinity stress. Similar findings reported in different crops such as canola (B. napus) (Ashraf and Ali, 2008), pea (Hemantaranjan *et al.*, 2014; Singh *et al.*, 2015), radish (Chaparzedah *et al.*, 2015), strawberry (Feghih *et al.*, 2017; Jamali *et al.*, 2016) and mustard (Husen *et al.*, 2018).

Plants have developed a complex antioxidative defense system, including antioxidant molecules and various antioxidative enzymes namely superoxide dismutase (SOD), catalase (CAT), ascorbate peroxidase (APX), guaiacol peroxidase (POD) and glutathione reductase (GR), which enable plants to perform better under saline environments. Salicylic acid (SA) act as a potential non-enzymatic antioxidant as well as a plant growth regulator, which plays an important role in regulating a number of plant physiological processes including photosynthesis (Fariduddin *et al.*, 2003; Singh and Usha, 2003; Waseem *et al.*, 2006; Arfan *et al.*, 2007). Several literature reported the ameliorative role of exogenous applied SA on redox homeostasis and membrane functioning. However the literatures are limited in regarding to the ameliorative role of exogenous applied SA on legume crops under salinity stress. Exogenous SA can regulate the activities of antioxidant enzymes and increase plant tolerance to abiotic stress (Li *et al.*, 1998; He *et al.*, 2002).

Various literatures suggested that effectiveness of SA in inducing stress tolerance depends upon type of species or concentration of SA applied (Borsani *et al.*, 2001; Németh *et al.*, 2002; Waseem *et al.*, 2006; Arfan *et al.*, 2007). Improvement or modification of plant growth and development can occur by the direct application of SA to seeds (Arfan *et al.*, 2007; Azooz and Youssef, 2010). Yang *et al.* (2004) reported that high endogenous SA level in rice acted as an antioxidant to scavenge ROS and/or SA indirectly modulated redox balance through activation of antioxidant pathways to protect cells from oxidative damage caused by ROS. Mandhania *et al.* (2006) concluded that under salinity stress, the stimulation of APX and GR activity was much higher in salt-tolerant than those of salt-sensitive cultivar. He *et al.* (2008) conducted a experiment in green house in which tomato (*Lycopersicon esculentum* L. cv. Hezuo 903) seeds were germinated on moist filter paper in dark at 28 °C for 3 d, and then they were grown in vermiculite for 20 d. The seedlings were transplanted into plastic containers. The seedlings were cultivated in a growth chamber (relative humidity 70 per cent, day/night temperature 28/20 °C, irradiance 800 µmol m^{-2} s^{-1} and 12-h photoperiod) for 10 d, and then pre-treated with a foliar spray of 1 mM SA. Salt treatment (NaCl @ 100 mM) started by adding NaCl to the nutrient solution. Obtained results showed that exogenous application of SA enhanced the activity of SOD, APX and CAT in NaCl stressed tomato plants. Azooz (2009) observed the differential responses of two faba bean (Vicia faba L.) local Egyptian genotypes to salinity (0 or 140 mM NaCl) and seed priming with 0.2 mM salicylic acid. The results indicated that both faba bean genotypes can develop different mechanisms of adaptation to salt stress and SA treatment enhances the antioxidant activity (CAT, APX, POX, GR) under salt stress, which were more in genotype 115 than genotype 125 over the control plants. This result provides considerable reason that the genotype 115 could be more salt-tolerant than genotype 125. Similar results were reported in spring wheat by Arfan (2009). Barba-Espin *et al.* (2011) reported that SA (50–100) decreased APX (an H2O2-removing enzyme) in pea plants and increased SOD (an H2O2-producing enzyme) activity under salinity stress.

Aftab *et al.* (2011) found that when *Artemisia annua* L. seeds were sown at a depth of 2 cm and salinity treatment was applied when plants attained three to four true leaf stages. To avoid osmotic shock, the NaCl concentration was increased gradually using 25 mM of NaCl daily until a final concentration of 50, 100, and 200 mM was maintained. A foliar spray of SA (@ 1.00 mM) was applied. Results indicated that SA enhanced the activities of the antioxidant enzymes CAT, POX, and SOD in salt-affected plants, with the highest activities of POX and SOD observed at 200 mM NaCl + 1.00 mM SA, whereas CAT activity was the highest at 50 mM NaCl + 1.00 mM SA. Maximum ROS generation occurred with 100 mM NaCl +1.00 mM SA, decreasing slightly when a higher concentration of salt was applied to the soil. The application of SA have increased the activity of anti-oxidant enzymes namely SOD, APX, GR, (Orabi *et al.*, 2013; Faghih *et al.*, 2017; Jamali *et al.*, 2016; Ahmad *et al.*, 2020), POX, CAT (Husen *et al.*, 2018), ascorbate oxidase (Behairy *et al.*, 2017). Besides strengthening the plant anti-oxidant system, SA also have a role in lowering the production of ROS such as MDA (Kumara *et al.*, 2010; Hemantaranjan *et al.*, 2014), H2O2 and superoxide anion (Singh *et al.*, 2015; Chaparzedah *et al.*, 2015; Ma *et al.*, 2017).

In addition to the antioxidant enzymes, SA also favoured the polyphenol (Jamali *et al.*, 2013) and flavonoids accumulation (Chaparzedah *et al.*, 2015) that in turn gives the plant a tolerance against biotic and abiotic stresses. It was reported by Shareef (2019) that the application of 500 ppm SA to the date palm plants increased the ascorbic acid content, which is a potential antioxidant.

D. Yield and Yield Attributs under Salinity Stress as Influenced by Salicylic Acid (SA)

The soil salinity affects large areas of the world's cultivated land, causing significant reductions in crop yield (Grewal, 2010). The soil salinity is one of the most problem affecting agriculture productions in the world an estimated 6 per cent of total global land area, which equates to between 400 and 950 × 106 hectares and nearly half of all irrigated land being adversely affected by high salinity. High salt stress disrupts the homeostasis of water potential and ion distribution at both the cellular and the whole-plant levels (Wenxue, 2003). Deleterious effects of salinity on plant growth and development is directly related to plant productivity (total plant biomass accumulation and grain yield).

The reduction in crop yield under saline environmental condition has been reported by several other workers, *e.g.*, in in pea (Bishnoi *et al.*, 1991). Shahid *et al.* (2011) observed that when NaCl (1 and 10 mM) applied alone in pea genotypes, reduced yield attributes but the numbers of pods per plant remain statistically non-significant. Among the NaCl treatments, 10 mM was more deleterious than 1 mM NaCl treatment.

Thus, salicylic acid conspicuously plays a role in enhancing the yield and yield components in many crops. Exogenous application of salicylic acid (SA) to different crops has been shown to elicit yield and yield components. An increase in number of pods and yield has been found in in Brassica juncea (Fariduddin *et al.*, 2003). According to Senaratna *et al.* (2003), SA is effective in inducing tolerance to heat, drought and chilling stress in bean and tomato plants. Murtaza *et al.* (2007) observed maximum increase in all the yield and yield components in pea at SA concentration of 10-4 M over SA 10-5 M and SA 0 M.

Farouk *et al.* (2011) suggested that depressive effects of sodium chloride salt stress on pea growth and yield can be alleviated and/or modified to some extent by presoaking seeds and spraying plants with appropriate concentrations of salicylic acid especially at 50 and 100 mg L^{-1}, calcium chloride at 2.5 and 5 mM or thiamin at 25 and 50 mg L^{-1}. Subsequently, Gad El-Hak *et al.* (2012) conducted an experiment to study the beneficial effects of foliar application with some antioxidants, *i.e.*, ascorbic acid and salicylic acid at the concentrations of 0.0, 200 ppm and mixture of both of them at 100 ppm along with humic acid at the rate of 0.0, 1.0 and 2.0g L^{-1} on plant growth, green pods, seed yield, yield components, quality and seeds chemical composition of pea cv. Master. The effect of foliar application with SA (200 ppm) significantly increased pod diameter, fresh seeds weight pod-1, number of fresh seeds pod-1 and green pod yield coupled with heaviest seeds showed top dry seed yield and 1000-seed weight.

When the plant of common bean received a foliar spray of 0.1 mM SA produced higher pod and seed yield, under induced salinity stress (Rady *et al.*, 2014). Mirdad and other (2014) witnessed enhanced yield in broccoli through SA application by means of improved main and secondary curd yield, number of curd plant^{-1}, curd diameter, curd weight and curd dry matter. With the sea water treatment as source of salinity stress, Orabi *et al.* (2013) found that the SA could improve the effect of salt stress on yield through increased pod weight, seed weight, no seeds plant^{-1}, 1000 seed weight of *Vicia faba*. The experimental results of Ahmad *et al.* (2020) showed that the seed priming with 0.2 mM SA improved flower and fruits number per plant and yield in sweet pepper. The acetyl salicylic acid also provided an effect as pronounced by the SA in tomato, through increased cluster number, flower number, fruit weight and yield (Sajyan *et al.*, 2019).

In strawberry cv. Festival, a foliar spray of 3.00 mM SA resulted in earliness, increased number of flowering cluster and total fruit yield (Mohamed *et al.*, 2017). Similarly, Shahmoradi *et al.* (2018) realized yield maximization in jasmine with SA application attributed to improved flower number and flower diameter.

The aforementioned results of diverse studies made with different plant species signify the protective role of salicylic acid under salinity stress, which logically suggests to extend abiotic stress research with special reference to salinity stress in different crops, for a meaningful exploitation of salicylic acid as a potential growth regulator for better survival of the plants especially under salinity stress. The salinity stress has been observed to be widespread with every increasing day and, thus, needs vital efforts from all spheres for its gradual eradication in soil and in plants of economic significance so as to contribute for holistic developments in the science of agriculture.

Pathogenecity

The effect of salicylic acid (SA) on plant resistance against environmental stress factors. Remarkably, amongst growth regulators, salicylic acid has been identified as a key signal for the expression of pathogenesis-related proteins. It is a fundamental requirement of a successful acclimation to control the oxidative chain-reactions and/ or maintain effective repair and detoxifying mechanisms. Certain convincing facts in this connection are described hereunder. Salicylic acid (SA) is an endogenous growth regulator of phenolic nature, which participates in the regulation of physiological processes in plants such as growth, photosynthesis, nitrate metabolism, ethylene production, heat production and flowering (Hayat *et al.*, 2010) and also provides protection against biotic and abiotic stresses such as salinity (Kaya *et al.*, 2002). SA induces acidic pathogen-related (PR) genes and inhibits basic PR genes, whereas jasmonic acid (JA) does the opposite (Wang *et al.*, 2001). SA also reduced the synthesis of tomato proteinase inhibitors. Antagonistic interactions between SA and JA affect the expression of PR protein genes in tomato (Thaler *et al.*, 1999).

REFERENCES

Ahmad F, Singh A and Kamal A. 2017. Ameliorative effect of salicylic acid in salinity stressed *Pisum sativum* by improving growth parameters, activating photosynthesis and enhancing antioxidant defense system. Biosci. Biotech. Res. Comm. 10(3): 481-489.

Ahmed WM, Imran M, Yaseen T, Haq MU, Jamshaid SR, Ikram RM, Ali A, Ali MM, Arif M and Khan A. 2020. Role of salicylic acid in regulating ethylene and physiological characteristics for alleviating salinity stress on germination, growth and yield of sweet pepper. Peer J 8: e8475 http: //doi.org/10.7717/peerj.8475

Azamal Husen, Iqbal M, Sohrab SS and Kafeel M and Ansari A. 2018. Salicylic acid alleviates salinitycaused damage to foliar functions, plant growth and antioxidant system in Ethiopian mustard (*Brassica carinata* A. Br.), Agric and Food Secur. 7: 44 https: //doi.org/10.1186/s40066-018-0194-0

Bayat H, Alirezaie M and Neamati H. 2012. Impact of exogenous salicylic acid on growth and ornamental characteristics of calendula (*Calendula officinalis* L.) under salinity stress. *Journal of Stress Physiology and Biochemistry*, 8(1): 258-267.

Behairy RT, El-Hamamsy SMA and El-khamissi HAZ 2017. Alleviation of Salinity Stress on Fenugreek Seedling Growth Using Salicylic acid, Citric acid and Proline. *Middle East J. Agric. Res.*, 6(2): 474-483.

Chaparzadeh N and Hosseinzad-Behboud. 2015. Evidence for Enhancement of Salinity Induced Oxidative Damages by Salicylic Acid in Radish (*Raphanus sativus* L.). Journal of Plant Physiology and Breeding. 5(1): 23-33.

Dong C, Wang X and Shang Q. 2011. Salicylic acid regulates sugar metabolism that confers tolerance to salinity stress in cucumber seedlings. *Scientia Horticulturae,* 129: 629–636.

Esan AM and Olaiya CO. 2016. Effect of salicylic acid (SA) seeds soaking on the NaCl salt stress induced changes in soluble sugar and protein accumulation in organs of two genotypes of okra plants. 10(6): 105-110. https: //doi.org/10.5897/AJPS2016.1410

Eshghi S and Taffazoli E. 2013. Vegetative growth, Yield, fruit quality and fruit and leaf composition of strawberry cv. 'pajaro' as Influenced by salicylic acid and nickel sprays, journal of plant nutrition, 36: 7, 1043-1055, https: //doi.org/10.1080/01904167.2013.766803

Faghih S, Ghobad C and Zarei A. 2017. Response of Strawberry Plant cv. 'Camarosa' to Salicylic Acid and Methyl Jasmonate Application Under Salt Stress Condition. J Plant Growth Regul (2017) 36: 651–659. https: //doi.org/10.1007/s00344-017-9666-x

Faghih S, Ghobadi C and Zarei A. 2017. Response of Strawberry Plant cv. 'Camarosa' to Salicylic Acid and Methyl Jasmonate Application Under Salt Stress Condition. *J Plant Growth Regul* (2017) 36: 651–659, https: //doi.org/10.1007/s00344-017-9666-x

Hadi H, Najafabadi A and Amirnia R. 2014. Comparison of different treatment methods of salicylic acid on some physiological traits of white bean under salinity stress. Cercetări Agronomice în Moldova, 3 (159): 97-105. https: //doi.org/10.2478/cerce-2014-0030

Hemantaranjan A, Singh R, Patel DK, Yadav A and Nishant Bhanu. 2014. Impact of salicylic acid on salinity tolerance in field pea (Pisum sativum L.) through intensified antioxidant defense system, physiological and biochemical attributes. Agrotechnol 2014, 2: 4, http: //dx.doi.org/10.4172/2168-9881.S1.011

Husen A, Iqbal M, Sohrab SS and Ansari MKA. 2018. Salicylic acid alleviates salinitycaused damage to foliar functions, plant growth and antioxidant system in Ethiopian mustard (*Brassica carinata* A. Br.). Agric and Food Secur (2018) 7: 44. https: //doi.org/10.1186/s40066-018-0194-0

Hussein MM and Abou-Baker NH. 2014. Growth and Mineral Status of Moringa Plants as Affected by Silicate and Salicylic Acid under Salt Stress. International Journal of Plant and Soil Science 3(2): 163-177, 2014; Article no. IJPSS.2014.004.

Jamalia B, Eshghia S and Kholdebarin B. 2016. Antioxidant responses of 'Selva' strawberry as affected by salicylic acid under salt stress. *Journal of Berry Research* 6: 291–301. https: //doi.org/10.3233/JBR-160130

Khan HA, Pervez MA, Ayub CM, Ziaf K, Balal RM, Shahid MA and Akhtar N. 2009. Hormonal priming alleviates salt stress in hot Pepper (*Capsicum annuum* L.). *Soil and Environ*. 28(2): 130-135.

Khan HA, Pervez MA, Ayub CM, Ziaf K, Balal RM, Shahid MA and Akhtar N. 2009. Hormonal priming alleviates salt stress in hot pepper (*Capsicum annuum* L.). Soil and Environ. 28(2): 130-135.

Ma X, Zheng J, Zhang X, Hu Q and Qian R. 2017. Salicylic Acid Alleviates the Adverse Effects of Salt Stress on *Dianthus superbus* (Caryophyllaceae) by Activating Photosynthesis, Protecting Morphological Structure, and Enhancing the Antioxidant System. Frontiers in Plant Science. 8(600), https: //doi.org/10.3389/fpls.2017.00600

Mirdad ZM 2014. Effect of K+ and Salicylic Acid on Broccoli (*Brassica oleraceae* var. *Italica*) Plants Grown Under Saline Water Irrigation. *Journal of Agricultural Science*, 6(10): 57-66.

Mohamed RA, Al-Kharpotly Abdelbaset and Abd-Elkader DY. 2017. Salicylic Acid Effects on Growth, Yield, and Fruit Quality of Strawberry Cultivars. *Journal of Medicinally Active Plants* 6, (2): 1-11. https: //scholarworks.umass.edu/jmap/vol6/iss2/2

Orabi SA, Mekki BB and Sharara FA. 2013. Alleviation of Adverse Effects of Salt Stress on Faba Bean (*Vicia faba* L.) Plants by Exogenous Application of Salicylic Acid. *World Applied Sciences Journal* 27 (4): 418-427. https: //doi.org/10.5829/idosi.wasj.2013.27.04.13653

Rady M, Mohamed GF, Abdalla AM and Ahmed YHM. 2015. Integrated Application of Salicylic Acid and *Moringa Oleifera* Leaf Extract Alleviates the Salt-Induced Adverse Effects in Common Bean Plants. *International Journal of Agricultural Technology* 2015 Vol. 11(7): 1613-1633.

Raghami M, Estaji A, Bagheri V and Aryakia E. 2016. Effect of salinity stress and salicylic acid on some morphophysiological characteristics of eggplant (*Solanum melongena* var. Taki) in soilless culture. *J. Sci. and Technol. Greenhouse Culture,* 7(27): 87.

Sajyan TK, Chokor M, Shaban N and Sassine YN. 2019. Enhancing salt tolerance of tomato (*Solanum lycopersicum*) by foliar application of aspirin (acetyl salicylic acid). Acta Hortic. 1253. XXX IHC – Proc. Int. Symp. on Water and Nutrient Relations and Mgt. of Hort. Crops. https: //doi.org/10.17660/ ActaHortic.2019.1253.7

Shahmoradi H and Naderi D 2018. Improving Effects of Salicylic Acid on Morphological, Physiological and Biochemical Responses of Salt-imposed Winter Jasmine. *International Journal of Horticultural Science and Technology,* 5(2): 219-230.

Shareef HJ 2019. Salicylic acid and potassium nitrate promote flowering through modulating the hormonal levels and protein pattern of date palm *Phoenix dactylifera* 'Sayer' offshoot Acta agriculturae Slovenica, 114/2, 231–238.

Singh AK and Dwivedi P. 2018. Modulating effect of Salicylic acid and Trichoderma in pea (*Pisum sativum* L.) under salt stress. *Journal of Pharmacognosy and Phytochemistry* 2018; 7(2): 2576-2582.

Souri MK and Tohidloo G. 2019. Effectiveness of different methods of salicylic acid application on growth characteristics of tomato seedlings under salinity. *Chem. Biol. Technol. Agric.,* 6: 26. https: //doi.org/10.1186/s40538-019-0169-9

Sunaina and Singh NB. 2014. 'Effect of salicylic acid on cabbage (*Brassica oleracea* var. *Capitata*) grown under salinity stress'. *Iranian Journal of Plant Physiology,* 4 (4), 1109-1118.

Yildirim E, Turan M and Guvenc I. 2008. Effect of Foliar Salicylic Acid Applications on Growth, Chlorophyll, and Mineral Content of Cucumber Grown Under Salt Stress, *Journal of Plant Nutrition,* 31: 3, 593-612.

Chapter 10

Technologies to Mitigate Climate Change for Sustainable Vegetable Production

Kanchan Bhan[1] *and Manish Bhan*[2]

[1]*Vegetable Breeder, Department of Horticulture,*
[2]*Agrometeorologist, Department of Physics and Agrometeorology,*
JNKVV, Jabalpur, Madhya Pradesh

Climate change resulted in a change of mean of weather parameters such as temperature, rainfall, relative humidity, wind speed, atmospheric gas composition, *etc.* for a long time in a large geographical area. It can be referred that, changes over time may be due to natural variability, or as a result of human activity. With the onset of industrial revolution, changes in climate have been observed due to changes in greenhouse gases (CO_2, N_2O, CH_4) emissions (GHGs) that cause to absorb more infrared radiation reflected from the earth's surface, which may alter energy balance of our climate system, causing the atmosphere to warm (IPCC, 2013). The changes in climate include variations in temperature, increase in soil salinity, water logging, high atmospheric GHGs concentrations and ultra-violet radiation. Global combined surface temperatures over land and sea have been increased from 13.6 °C in 1881-90 to 14.47 °C in 2001-10 (WMO, 2013). Since 1901 year, the mean annual temperature of India is increased by 0.46 RC (Data Portal India, 2013; Ayyagari *et al.*, 2014). The temperature increase will alter timing and amount of rainfall, availability of water, wind pattern and causes incidences of weather extremes events (Minaxi *et al.*, 2011). Additionally, rising temperature also enhances the incidences of insect-pests and diseases among agricultural crops and they render their cultivation as unprofitable (Ayyagari *et al.*, 2014). Similarly, sowing of potato at a new optimal date of mid-November will reduce yield losses by 8 per cent caused due to weather extremes (Luck *et al.*, 2010), as potato requires long days with low temperature for its flowering stage. Due to increase in temperature, the potato breeding area is shifting towards the further more altitude. Similarly, high temperature in tomato affects fruit colour as lycopene pigment degrades at high temperature above 27 RC and destroyed at 40 RC. High temperature effect pollination and fruit set in tomato (Kalloo *et al.*, 2001). In chilli, high temperature affects post-pollination stage; inhibit fruit set and red colour development (Erickson and Markhart, 2002). Similarly, decrease in minimum temperature affects seed germination in chilli (Flynn *et al.*, 2002).

Vegetables are an important component of human diet as they are the only source of nutrients, vitamins and minerals. Vegetables play a major role in Indian agriculture as they ensure food and nutritional security of the country apart from enhancing per capita income of the farmers. They are good remunerative to the farmer as they fetch higher price in the market. Considering the recommendations of Indian Council of Medical Research for nutrient requirement, vegetables form an important component of a balanced diet, which remains a distant dream for many Indians. Indians are predominantly vegetarian and depend on vegetables and fruits for bulk of their nutrients and minerals requirement. Currently, India is producing 15 per cent of world vegetables and ranks second in vegetable production after China and contribute 16.7 per cent to global vegetable area and 15.4 per cent to production. In India, production increased from 58.5 mt in 1991-92 to 146. 5 mt in 2010-11 (Kumar *et al.*, 2011).But, the productivity is just 17.4 tonnes per ha which is less than the world average of 18.8 tonnes per ha. One of the reasons for this low productivity is low access of Indian farmers to the modern technologies as per the changing weather conditions.

Figure 10.1: Tomato Plant Burned Due to Extreme Low Night Temperature or Frost at Jabalpur (*Picture Courtesy*: Dr. Manish Bhan).

Figure 10.2: Chilli Plants sever Injury Due to Extreme Low Night Temperature or Frost at Jabalpur (*Picture Courtesy*: Dr. Manish Bhan).

Vegetables supplies vitamins and minerals apart from major nutrients, and supposed to be a best resource for overcoming micronutrient deficiencies. Vegetable production sector directly dependent on climate and weather conditions and its study will be a priority to alleviate the adverse impact of climate change. The majority of climate change studies are on forest, cereals and fruit trees but rarely on short-lived vegetables (Bisbis *et al.*, 2018). They are sensitive to environmental extremes like high temperature and limited moisture supply that affect yield levels apart from several physiological and biochemical processes like reduced photosynthetic activity, altered metabolism and enzymatic activity, thermal injury to the tissues, reduced pollination, and fruit set, *etc.* (Ayyagari *et al.*, 2014).

Concerns

- ☆ Water stress for irrigation
- ☆ Not sufficient storage and cold store facilities
- ☆ Weak Market linkages
- ☆ Less practice of agro –horticultural systems
- ☆ Increased incidences of pest attacks
- ☆ Increasing temperature affecting the pollen viability and crop growth
- ☆ Frequent unpredicted severe events affecting the crop produce
- ☆ Soil moisture stress to the crops leading to their stunted growth
- ☆ Decreased seed production due to increasing temperature.

Source: SKMCCC, MPSAPCC. 2014.

Madhya Pradesh is located in Central India having 51 districts and 48 KVKs under 11 different agro-climatic zones in the state. The economy of the state is predominantly based on rainfed agriculture (only 1/3rd of grossed cropped area is being irrigated) and the major crops are paddy, wheat, maize and sorghum among cereals. Similarly, chickpea, pigeonpea, blackgram, and greengram among pulses; soybean, groundnut and mustard among oilseeds; and cotton and sugarcane among commercial crops. Horticultural crops like potato, onion, garlic, along with fruits like papaya, banana, oranges, mango and grapes are also grown in the state. The major constraints in productivity enhancement are non-availability of suitable high yielding varieties; heavy weed infestation, high incidence of disease and pests, lack of technological interventions, crop damage due to erratic rainfall, and lack of awareness about production technology (ICAR, 2019).

The challenges that climate change brings along will require adaptation strategies in order to meet consumer demands and to ensure food security standards. Special efforts need to be taken to develop state specific strategy with major focus on the key agricultural technologies and their combinations along with effective technology delivery mechanism and market linkage. This not only alleviate climate change drawbacks but also focus on the plans for enhancing production, cost reduction, quality improvement and generating additional income.

Challenges due to Climate Change on Vegetable Production in Madhya Pradesh

Vegetable crops bring in the income to the small and marginal farmers due to high productivity and high value of produce as compared to foodgrains. These crops are beneficial to the farmers as grown with minimum farm inputs. There are certain challenges that need to overcome to enhance its production under climate change scenario. Of the major constraints, water availability is the major challenge for vegetable production in the state. Their requirement will increase in the future due to rise in the temperatures thereby affecting production. Beside this, rise in temperature leads to early maturity and fruit fall among the temperature sensitive vegetables. It also reduce time period of tuber initiation process especially in potato, or affecting colour quality in tomato, or lead to decrease in pollen viability in many crops. It may also lead to bolting in crucifers whereas anthocyanin production is affected in chilli crop. With the rise in temperatures, photoperiod may not show variation. A photosensitive crop like onion matures faster leading to small bulb size. Similarly, rise in temperatures during winter will affect vegetable maturity period. The effect of high temperature effect on the quality of different vegetables is given in the Table 10.1.

Additionally, extreme weather events like hailstorm, frost (chilling injury) and drought conditions affect produce badly that may give rise to the incidence of insect-pest and disease infestation. Apart from this, state lacks market linkage, and short shelf-life of the produce create risk that makes less lucrative for growing vegetables in the state.

Table 10.1: Effect of High Temperature Stress on different Qualities and Nutrition in Vegetables (As copied from Bisbis *et al.,* 2018)

Qualities Affected in Vegetables	*Observed in the following Vegetables*	*Reference*
External quality		
Low tip fill of the cob	Sweet corn	Wien (1997)
Increase in tip burn	Lettuce, broccoli, chinese cabbage	Sauré (1998); Gruda and Tanny (2014)
Decrease in seed and fruit size	Bean, pea, tomato	Siddique and Goodwin (1980); Lattauschke (2015); Gruda (2005)
Increase in fibre content	Bean	Peet and Wolfe(2000)
Increase in flower bud size	Broccoli	KalauŸewicz *et al.* (2009)
Increase in loose heads	Broccoli, lettuce	KalauŸewicz *et al.* (2009); Wien (1997)
Increase in bulb splitting	Onion	Peet and Wolfe (2000)
Decrease in fruit coloration	Tomato	Gruda and Tanny (2014)
Increase in fruit cracking	Tomato, pepper	Gruda (2005); Rosales *et al.* (2010): Gruda and Tanny (2014)
Increase in sunburn	Pea, tomato, pepper	Rosales *et al.* (2010): Gruda and Tanny (2014)
Internal quality		
Decrease in sugar content	Pea, tomato, cabbage, melon, sweet corn	Rosales *et al.* (2010); Gruda and Tanny2014; Wien (1997)
Decrease in starch content	Sweet com	Wang and Frei (2011)
Decrease in Macronutrients (K, Mg, Ca)	Tomato	Rosales *et al.* (2010)
No change in micro-nutrients (Fe, Zn, M, Cu)	Tomato	Rosales *et al.* (2010)
Increase in Ascorbic acid (Vitamin C) content	Tomato, lettuce	Rosales *et al.* (2010)
Decrease in Ascorbic acid (Vitamin C) content	Tomato	Wang and Frei (2011)
No change in Ascorbic acid (Vitamin C) content	Broccoli	Mølmann *et al.* (2015)
Increase in Tocopherol (Vitamin E)	Lettuce	Wang and Frei (2011)
Decrease in Lycopene content	Tomato	Gruda (2005); Rosales *et al.* (2010)
Decrease in Carotene content	Carrot, tomato, lettuce	Ibrahim *et al.* (2006); Rosales *et al.* (2010); Wang and Frei (2011)
Increase in antioxidants (anthocyanin, flavonols, phenols, glycosinolates)	Tomato, *Brassica* species *e.g.* broccoli	Rosales *et al.* (2010); Gruda (2005); Mølmann *et al.* (2015)
Increase in Organic Sulfur (flavor)	Onion	Coolong and Randle (2003)
Increase in Terpenes	Carrot	Ibrahim *et al.* (2006)
Increase in bitter compounds	Lettuce	Wien (1997)

Climatic Current Status and Projections of Madhya Pradesh

Different weather parameters affect crop growth and yield status. With the rapid change in weather conditions in the state, crop quality and quantity is affected including vegetables. This will enhance further in the coming years due to climate change scenario. The rainfall and minimum temperature trends of the state are depicted in the Figure 10.1 (As copied from Venkateswarlu and Rao, 2015 under AICRP on Agrometeorology, ICAR-CRIDA, Hyderabad). The districtwise historical rainfall data for the state showed divergent trends as depicted in the Figure 10.3.

No significant trends in annual rainfall observed in majority of the districts however, districts in Central and north-east parts of the state observed a decreasing trend. Among the seasons, decreasing trend in rainfall during June – September (*Kharif* season) months among major tracts of Central area and fewer districts under north-west area of the state (Figure 10.3). Similarly, *rabi* season (November-March) months did not showed major changes in rainfall, except few districts in south-east area of the state. There is an increase in annual, maximum and minimum

Figure 10.3: Significant Trends of Rainfall (1971-2011) and Minimum Temperature (1971-2012) as Annual and Season-wise in Central Indian State of Madhya Pradesh.

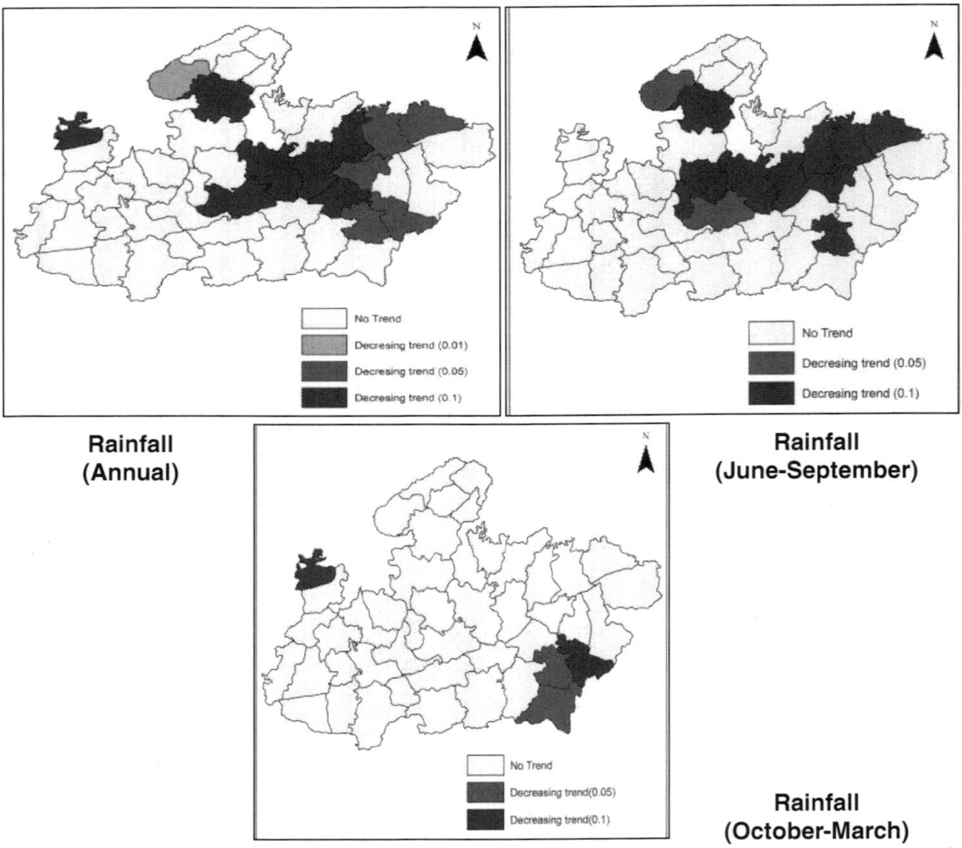

Rainfall (Annual)

Rainfall (June-September)

Rainfall (October-March)

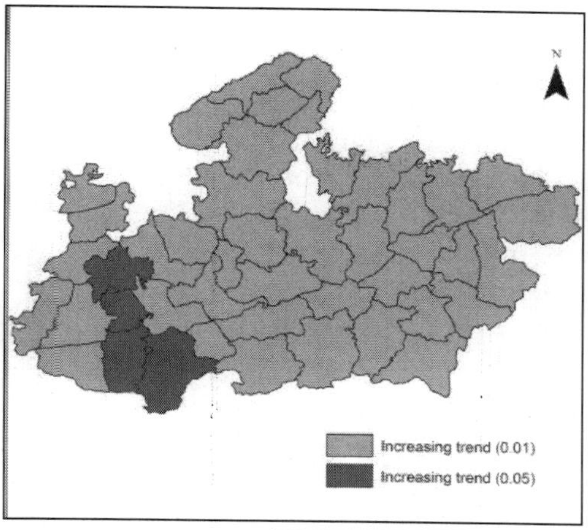

Minimum Temperature (Annual)

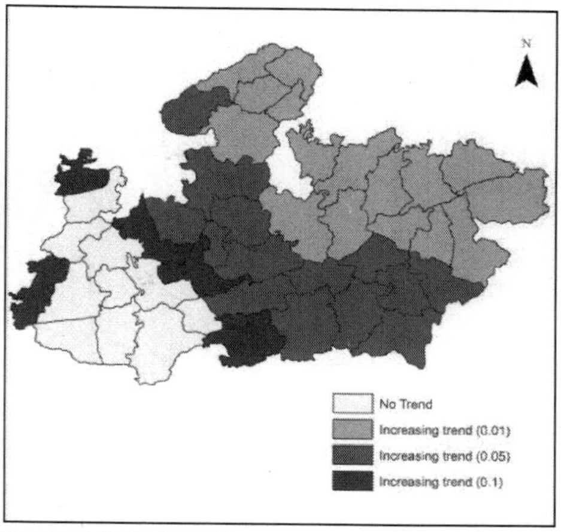

Minimum Temperature (June-September)

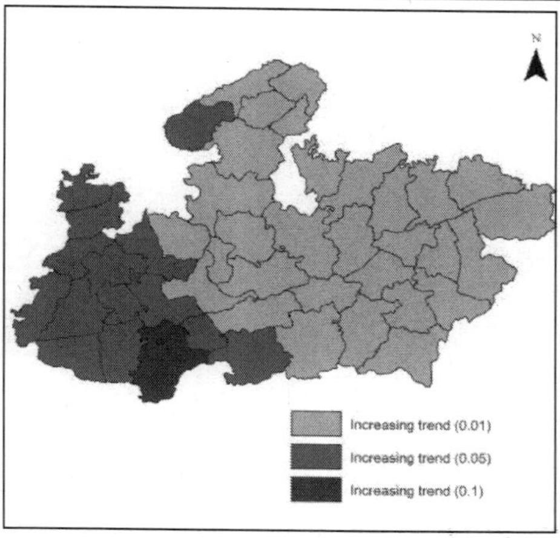

Minimum Temperature (October-March)

(*Data source*: Rainfall; IMD district wise data;
Temperature: 0.5 degree CRU monthly data converted into area weighted district)
(*Source*:Venkateswarlu and Rao, 2015).

temperatures by 0.6, 0.6 and 0.62 RC in the last 102 years (1901-2002) in Central India with warming more pronounced in winter than summer seasons (Duhan *et al.*, 2013). A rise in day temperature leads to more evapotranspiration whereas rise in night temperature affects reproductive performances of the crops especially during grain filling period (Venkateswarlu and Shanker, 2009). Among the vegetables, high temperature reduced crop development stages in common bean, pea, lettuce, onion however it also increase growth duration in cauliflower and broccoli vegetables (Bisbis *et al.*, 2018).

Climate Project of Madhya Pradesh

Among the climatic change projections of Madhya Pradesh using PECIS-A1B scenario for 2050's year suggest that the state receive relatively more rainfall in the future as compared to current scenario however, it will be in the form of intense storms that create implications for crop production, cause physical damage to the fruits thereby impact quality of the produce, and affect water resource management in the state.

As per the work did by the Climate change Cell of State Knowledge Management on Climate Change (SKMCC) under Environmental Planning and Coordination Organization Bhopal (EPCO), datasets of 2030s (2021-2050) and 2080s (2071-2098) was derived with a grid resolution of 0.44° x 0.44° using PRECIS (Providing Regional Climate for Impact Studies) model were developed for climate projections of Madhya Pradesh. As per their studies, the daily maximum temperature may rise between 1.8-2.0 °C and minimum temperature between 2.0-4.0 °C in the period 2030s with the eastern part experiences more warning then the western part of the state. By 2080s, maximum temperature may rise between 3.4 – 4.4 °C with more probability in the northern region, while minimum temperature increase up to 4.4 °C than normal in all the areas of Madhya Pradesh.

As per rainfall projections from the same agency, decrease in winter rainfall may occurred and moves from eastern to western part of Madhya Pradesh for the period from 2021 to 2050 years. There is a slight increase in rainfall during south-west monsoon period in all over the state, while pre-and post-monsoon periods have variability as mentioned in the Table 10.2.

Table 10.2: Projected Change in Climatic Parameters of Madhya Pradesh by 2100 using PRECIS model (Developed by GoMP, Climate Change Cell, EPCO, Bhopal)

Sl.No.	*Projected Change of Climate*	*2010-2050*	*2071-2100*
1.	Daily maximum temperature	1.8 – 2.0 °C increase	3.4 - 4.4 °C increase
2.	Daily minimum temperature	2.0 – 2.4 °C increase	> 4.4 °C increase
3.	Monsoon precipitation	Increase in precipitation by 1.25 times than current normal in most parts of MP. However, no change in Morena, Shivpuri, Gwalior and Bhind districts. Increase in Hoshangabad, Betul and southern part of Sehore districts.	Increase in 1.35 times from current normal, with majority in Hoshangabad, Damoh, Mandla and Balaghat districts.
4.	Winter precipitation	Decrease in precipitation	Substantial increase in precipitation by 1.45 -1.85 times in central and south-western parts of MP.

As per the future climatic projections, precipitation will increase along with temperatures (maximum and minimum) that will increase the rate of evapotranspiration. This causes more release of water vapours into the atmosphere that helps to increase directly the temperature of the atmosphere causing more global warning due to greenhouse effect than the current normal scenario. Based on these facts, climate change is already anticipated to create havoc globally thereby influencing eco-systems as well as agriculture. Strategies need to be developed to alleviate its effect in the coming future, particularly for Central Indian conditions.

Physiological Response of Vegetable due to Rise in Weather Parameters

The physiological processes in vegetables help in their metabolism and other growth stages in plant's life. These processes are highly affected by increasing temperatures, and CO_2 concentrations, and interactions between them. Response of weather parameters on plant growth are described briefly as:

(a) Elevated CO_2 Concentration

With the increase in CO_2 concentration in the atmosphere, C_3 plants stimulates more photosynthesis thereby exhibiting that climate change may contribute to higher yield in vegetable plants, as majority of vegetables are C_3 species. The higher CO_2 diffuses through the stomata and enter to the cell chloroplast, where *Rubisco* enzyme (involved in first step of carbon fixation) facilitates carboxylation, and reduce oxygenation hence photorespiration. It favours an increase in yield, and therefore, vegetables yield more under higher CO_2 concentration of 600-1000 ppm concentration. However, it is not always beneficial when the other factors like light, nutrients or water are in limited conditions (Reich *et al.*, 2015). In C_4 plants like maize, photosynthesis is thought to be CO_2 –saturated at current CO_2 concentrations, and thus not expected to benefit from changes in the rise of CO_2 concentrations.

(b) Elevated Air Temperature

Air temperature is a predominant environmental factor for vegetable growth and yield. Vegetable crops have species specific temperature requirements. The temperature optimum for hot season vegetables is between 25-27 °C, warm season crops is between 20-25 °C, and cold season crops is between 18-25 °C (Wein, 1997). Global warming is expected to increase both the maximum and minimum temperatures in future, thereby changing the growing conditions for vegetables. Plant response to elevated temperatures is not only species specific but also depends on growth stages and plant age. Generally, higher temperature promotes photorespiration however photosynthetic rate also showed positive response at higher optimum temperature. Elevated temperature benefit plants only as long as they do not exceed critical thresholds. In heat wave, where temperature exceed threshold value create stress on the physiological level thereby diminished yield and quality of the vegetables. Similarly, elevated temperatures also required sufficient water supply in the field to cool the plants through transpiration, thereby reduced water use efficiency. This is more common in C_4 plants however C_3 plants reduce losses by closing their stomata due to decrease in rate of photosynthesis at elevated temperature.

Overall, elevated CO_2 concentration in the vegetables can reduce heat stress with the conditions that ample water supply, light intensity, nutrient supply, *etc.* should be in good quality. If one of the latter is inadequate conditions, plant growth and yield will be in limited condition due to the deficit factor.

Apart from physiological response, growth habits in vegetables are generally classified as either determinate or indeterminate types, however there is variability among cultivars within species. The indeterminate crops appear less vulnerable due to the effect of high temperature as flowering is induced depending on photoperiod (*e.g.* climbing and staggering beans), or independent of both temperature and photoperiod (*e.g.* Tomato) (Wien, 1997; Bisbis *et al.*, 2018). In all, determinate vegetables might suffer from faster development due to climate change, possibly resulting in reduced yield and product quality. This could lead to medium-duration cultivar becoming early-duration cultivar, late- cultivar becoming medium-duration, and early –cultivar becoming increasing vulnerable (Keatinge *et al.*, 2012). In theory, indeterminate crops are expected to benefit from climate change, but further evidence has to be provided as experimental data suggest a different trend. Additionally, cultivar variations of the same plant species need further research on this aspect.

Strategies to Overcome Climate Change

1. Soil and water conservation through demonstration of best practices: The conservation technologies need to be adopted due to rising temperatures and CO_2 concentration with changes in rainfall pattern of the state. Promoting conservation techniques like mulching, drip irrigation, and micro irrigation practices. Creation of appropriate thermostatic conditions for horticultural crops like shed nets and polyhouses for supporting growth of vegetables should be encouraged.
2. Development of agri-horticultural system: Agri-horticultural systems needed to be upgraded for securing livelihoods, especially in tribal areas. Also encourages diversification through intercropping of field crops-vegetables, or crops-vegetables-fruit crops, *etc.* will help to boost cropping intensity thereby ensuring livelihood security.
3. Policies development: Developing of horticultural policies as per the recommendations of National Horticultural Mission (NHM) to create production storage facilities as per the agro-climatic conditions for promoting of planting vegetables according to the suitability of different agro-climatic zones.
4. Mechanization to produce and value-addition of products: Enhancing access to mechanization as well as value-addition of the products by creating market hubs, as vegetables have less shelf-life and subject to postharvest losses. Beside this, rise in temperature also increase humidity thereby affecting quality of the produce.
5. Research and extension support provided in the form of good cultivar seed and cropping techniques to the farmers to ensure productivity even in the climatic change context.
6. Creating cooperatives for enhancing the livelihood of small and marginal farmers. Strengthening of existing and formation of new marketing cooperatives may be considered as a potential option to boost its production.

Adaptation Strategy to Combat Climate Chnage in Vegetable Production

A possible adaptation strategy needs to be implemented to improve the sustainability in vegetable production. It includes securing water and nutrient supply while reducing damage caused by extreme weather events (like heat wave, cold wave, floods, drought) as well as adopting gene-pool of cultivars to the new growing conditions (Table 10.3). Implementation of adaptation strategies to benefit from the positive aspect of high CO_2 concentration that extend growth period for summer crops and cultivation of winter crops (Bisbis *et al.*, 2018).

Table 10.3: Climate Change Impact and Adapted Strategies to Alleviate Adverse Impacts (As copied from Bisbis *et al.*, 2018)

Extreme Weather Events	*Impact*	*Adaptation*	*References*
Extreme rainfall	Water logging thereby creating flood-like situation	Installation and maintenance of field drainage	Dietrich *et al.* (2015)
	Cause soil erosion	Increasing soil cover and organic matter by reduced tillage practice	Flaig (2013)
Summer drought	Reduced water availability and increasing irrigation demand	Mulching minimizes soil evaporation losses	Flaig (2013); De la Pea and Hughes (2007)
		Drip irrigation combined with tensiometers increases irrigation efficiency	Schneider (2016); Buttaro *et al.* (2015)
		Deep wells, bank filtrate, rainwater harvesting, aquifer recharge ponds, retention dam secure and store water resources	Dietrich *et al.* (2015); Moriondo *et al.* (2010)
Heat waves	Cause heat stress that resulted to loss in quality and quantity of the produce	Reduce solar radiation by shading devices	Flaig (2013); Shahak (2008)
		Increase heat and drought tolerance through grafting and breeding	Abdelmageed and Gruda (2009); De la Pea and Hughes (2007)
	Heat accumulation in greenhouses	Modernization of greenhouse constructions with improved ventilation and advanced climate control	Fink *et al.* (2009)
Pest and disease infestation	Increase outbreak of pests of fungi, weeds, insects and virus	Increased use of pesticides with respect to limit values and a broad range of active agents to avoid host resistance	Fink *et al.* (2009); Hullé *et al.* (2010)
		Weed suppressing soil cover and insect screens	Jamieson *et al.* (2012)
		Breeding for pest resistance	Mattos *et al.* (2014); Van de Perre *et al.* (2014).

The technologies as adopted by Bisbis *et al.* (2018) includes:

Integrated Water Management

Changes in the trends of rainfall along with an increase in their variability make water management a key factor in the challenges of climate change. In order to secure sufficient irrigation in agriculture, integrated ground- and surface water management is of utmost importance. In times of excess rainfall, the drained water should be stored for usage in times of scarcity. This water will later be helpful for earlier planting in spring and later planting in autumn become increasing important in order to profit from the excess rainfall thereby creating conditions for efficient crop production with less irrigation. On the other side, storms and hail can mechanically damage leaves and fruits of vegetables, with anti-hail nets being the common counter measure (Flaig, 2013). Heavy rainfall causes soil erosion, and thus a loss of fertile land. This can be avoided by reduced tillage practices, integration of cover- and catch crops to reduce fallowness, as well as improving water holding capacity in the soil by raising the content of organic matter (Ernst, 2012). Reduced tillage in vegetable production systems may increase soil organic matter, but needs to become more profitable for producers while fulfilling the requirements of both market and ecological issues.

Figure 10.4: Anti-hail Nets to Protect Vegetables from Hail.

Figure 10.5: Plastic Mulch for Growing Vegetables.

Similarly, aquifer recharge seems to be a suitable method to store surplus water, and can be used later during crop growing period. Furthermore, harvesting rain water into water reservoirs at times of intense rainfall might become essential for vegetable grower in dry regions of the state. Additionally, improve water efficiency reduce soil evaporation by covering the soil with mulches (organic or plastic or fleece mulch), or integrating catch crop and reduced tillage. Beside this, a wider use of drip irrigation and mulches to reduce soil evapotranspiration, and sensors measuring the soil volumetric water content, are suitable tools to manage plant irrigation and conserved water.

Figure 10.6: Drip Irrigation in Bell Pepper Plants. (*Picture courtesy*: Dr. Manish Bhan PhD Dissertation).

Figure 10.7: Land Laser Leveller Machine to Help Improve Water Use Efficiency in the Field.

Strategies to Adapt Heat Stress in Vegetables

Breeding heat and drought tolerant cultivars with a better water use efficiency as well as species specific traits such as secure vernalization for head formation *e.g.* in cauliflower, is of crucial importance (De la Pea and Hughes, 2007). A heat shock of young plants at high temperature and intense light may increase heat tolerance, and can make photosystem II more resilient towards heat exposure by producing heat shock proteins. However, more research is needed in the field of heat shock pre-treatment since results have shown no beneficial effects (Abdelmageed and Gruda, 2009).

Strategies Adopted to Manage Weeds, Insect-Pests and Diseases

Increase variation in temperatures and rainfall are expected to emerge weeds earlier while rising CO_2 concentration will increase their growth rate and established better than vegetables thereby competing better than the crop plants. Similarly, insect-pest mortality rate decline due to the short of cold season period, while increasing atmospheric CO_2 might alter the feeding quality of the host (Dáder *et al.*, 2016). Insect propagation is highly dependent on temperature and warming would enable them to produce more generations per year. This proliferation of sucking pests like aphids and whiteflies may increase spread of viruses in the future, as both are the effective vectors (Keatinge *et al.*, 2012). Beside this, new species may also emerge like South American leaf minor (*Tuta absoluta*) in tomatoes as a major pest in the European countries. Additional application of pesticide may be required that needed attention to limit residual values especially in fresh produce, with transgressions making the product unmarketable.

Other practices to be adopted include optimum time of sowing, crop diversity through crop rotations, organic farming that include practices of planting cover crops during fallow period, planting of living mulches as inter-row crops between the rows of vegetables, holistic cropping system of vegetables planted of different family; Use of weather based agromet advisories for weather projections in advance;

and postharvest management as state lacks in robust storage and processing facilities that pose threat to balance demand and supply of the produce.

Organic farms have greater diversity due to use of cultural practices like crop rotations with preference of cultivars that are tolerant to both biotic and abiotic factors of climatic extremes, insect-pests and diseases. They sustained diversity in the field by implementing holistic cropping systems that makes them flexile to respond well under climatic risk thereby suffers less damage compared to conventional farming systems that dominate with monoculture system. Additionally, organic farming preserve soil fertility and sustain organic matter to improve soil moisture capacity, or reduce the negative effect of drought. Hence, organic agriculture increases its ability to continue functioning even in unexpected events of climate change, and potentially contribute to mitigate threats from climate change on vegetable production.

Overall, sustainable alternatives might be the application of insect screen, weed suppressing mulch layers, and breeding of new cultivars not only with heat and drought tolerance, but also pest and disease resistance (Bisbis *et al.*, 2018).

Greenhouse Vegetable Production Adapted to Climate Change

The use of poly-tunnels or greenhouses might be a possible adaptation strategy to manage and conserve the resource water. Effective greenhouse production relies on environmental conditions. Greenhouse location should be carefully chosen with respect to the future changes, and older greenhouses need to be upgraded to higher technology standards. The site of the greenhouse determines framework conditions such as climate (irradiation, temperature, day length, *etc.*), edaphic (structure, chemical, and biological soil properties, soil water and air content at different water tensions) and management (cultivation measures) are the basic factors to be considered when constructing a greenhouse(Gruda, 2009). The increased problem with high humidity, as well as increased incidence and duration of heat waves in summer will require more effective climate control, which can only be achieved through innovative technologies. Shading, sufficient ventilation, and technical cooling might become indispensable in some locations during summer and dehumidification units during winter (Fink *et al.*, 2009).

Figure 10.8: Polyhouse are the Best Option for Controlling Weather Conditions for Growing Vegetables.

Conclusion

It is necessary to prioritize issues pertaining to farmers friendly climate resilient technologies under changes in rainfall distribution, temperature variation and increase in extreme weather events, *etc.* on the quality and quantity of vegetables. The future climate change modifications affects plant physiology of growing vegetables due to shifting of environmental factors. These physiological responses under climate change scenarios are still under-researched for different vegetables, and need work using simulated studies in which interactions between rising CO_2 levels with elevated temperatures and ozone enrichment needs to work out. Certain examples of climate change on vegetables as:

- ✰ Higher temperatures cause insufficient cold accumulation in some vegetables. Insufficient vernalization is unfavourable for head formation in cauliflower but may benefit lettuce and cabbage due to reduce incidence of bolting in spring season.
- ✰ In pea, onion and other determinant crops, yield reductions might emerge from hastened development at higher temperatures, resulting in reduced light interception during shortened vegetative stage.
- ✰ Heat wave will further reduce crop duration of certain vegetables and yield reductions in fruit crops. *e.g.* Beans might result from reduced fruit set.

In spite of this, choosing the right cultivar for the right location and season will become more significant for growers, while plant breeders will have to provide the industry with new cultivars to improve tolerance towards heat and drought stress, insufficient chilling and vernalization, hastened development, and with an extended shelf-life. Apart from it, pest and disease infestation will become more intense thereby need further research

as how climate change affect physiology and population dynamics of the pests, their interactions with plants under modified environment that helped to develop tolerant cultivars. Similarly in greenhouses, high relative humidity in winter creates problem whereas coping with excessive heat and irradiation in summer season is another problem. There is a broad range of technical tools that may be useful in dealing with these problems. Efficient water supply must be secured, and vegetable growers must come up with innovative solutions like drip irrigation, soil mulching, aquifer management and construction for stage of excess water reservoir.

Overall, it is anticipated that new adaptative techniques needs to be implemented to alleviate the affect of climatic change events. Implementing the innovative technologies could enable vegetable growers to mitigate the adverse effect of climate change and benefit from higher yield levels, and provide fresh and healthy vegetables for future generations.

REFERENCES

Abdelmageed AHA and Gruda N. 2009. Influence of grafting on growth, development and some physiological parameters of tomatoes under controlled heat stress conditions. *Eur. J. Hortic. Sci.* 16-20.

Ayyagari\K, Sidhya P and. Pandit MK 2014.Impact of climatechange on vegetable cultivation- A Review. *Intl. J. Agriculture, Environment and Biotechnology* 7: 145-155.

Bisbis MB, Gruda N and. Blanke M 2018.Potential impacts of climate change on vegetable production and product quality- A review. *J. Cleaner Production.* 170: 1602-1620.

Dáder B, Fereres A, Moreno A, and Trêbicki P P, 2016. Elevated CO_2 impacts bell pepper growth with consequences to *Myzus persicae* life history, feeding behaviour and virus transmission ability. Sci. Rep. 6.

Data Portal India, 2013. Annual and Seasonal Mean Temperature of India, National Informatics Centre of Govt. of India. Downloaded from http: //data.gov.in/dataset/annual-and- seasonal-mean-temperature-india

De la Pea R and Hughes J. 2007. Improving vegetable productivity in a variable and changing climate. *J. SAT Agric. Res.* 4 (1), 1-22.

Duhan D, Pandey A, Singh KP and Pandey RP. 2013. Spatial and temporal variability in maximum, minimum and mean air temperatures at Madhya Pradesh in central India. *C.R. GeoScience,* 345: 3-21.

Erickson AN and. Markhart AH 2002. Flower developmental stage and organ sensitivity of bell pepper (*Capsicum annuum* L.) to elevated temperature. *Plant Cell Environment* 25: 123-130.

Ernst M. 2012. Climate change and vegetable production in the upper Rhine rift. In: Der Oberrheingraben im Klimawandel eine Region passt sich an. Regionalkonferenz des Bundes und der Länder Baden-Württemberg, Hessen und Rheinland-Pfalz, pp. 51-52 (in German).

Fink, M, Kläring HP, and. George E 2009. Horticulture and climate change. In: Dirksmayer, W., Sourell, H. (Eds.), *Water in Horticulture,* Vol. 328. Landbauforschung, Sdh, pp. 1-9 (in German).

Flaig H. 2013. Adaptations to Climate Change in Baden-Würtemberg. Expert Assessment for Field of Action in Horticulture Part a. Landwirtschaftliches Technologiezentrum Augustenberg, Karlsruhe long version (in German).

Flynn R, Phillips R, Ulery A, Kochevar R, Liess L, and. Villa M. 2002. Chile Seed Germination as Affected by Temperature and Salinity, Report 2, New Mexico Chile Task Force pp: 1- 12.

Gruda N. 2009. Do soilless culture systems have an influence on product quality of vegetables? *J. Appl. Bot. Food Qual.* 82; 141-147.

ICAR. 2019. State specific strategies for doubling farmers income-2022. Indian Council of Agricultural Research, DARE, New Delhi. pp. 293-325.

Kalloo G., Benarjee MK, and Tiwari RN. 2001. Tomato. p. 10-28. In Thumbraj, S and N. Singh. [eds.] Vegetables Tuber crops and Spices. Directorate of Information and Publication in Agriculture, Indian Council of Agricultural Research, New Delhi.

Keatinge JDH, Ledesma DR, KeatingeFJD, and Hughes JA. 2012. Projecting annual air temperature changes to 2025 and beyond: implications for vegetable production worldwide. *J. Agric. Sci.* 152 (1); 38-57.

Kumar B., Mistry NC, Chander BS, and Gandhi P, 2011. Indian Horticulture Production at a Glance. Indian Horticulture Database-2011. National Horticulture Board, Ministry of Agriculture, Government of India.

Luck JM, Asaduzzaman S, Banerjee I, Bhattacharya K, Coughlan GC, Debnath D, De Boer, Dutta S, Forbes GW, Griffiths D, Hossain S, Huda R, JagannathanS, Khan G, O'Leary, Miah G, Saha A. Spooner-Hart R. 2010. Project report of Asia pacific net work for global change research entitled "The effects of climate change on pest and diseases major food crops in the Asia pacific region" Downloaded from http: //www.apngcr.org/newAPN/activities/ARCP/2010/ARCP2010_05CMY_Luck/ARCP2010-05CMY-Luck-FinalReport.pdf

Minaxi RP, Acharya KO and Nawale S 2011. Impact of Climate Change on Food Security. *International Journal of Agriculture, Environment and Biotechnology* 4(2): 125-127.

Reich M, Meerakker AN, Parmar S, Hawkesford MJ, and L.J. De Kok. 2015. Temperature determines size and direction of effects of elevated CO2 and nitrogen form on yield quantity and quality of Chinese cabbage. Plant Biol. 18 (S1); 63-75.

Stocker TF, Qin D, Plattner GK, Tignor M, Allen SK, Boschung J, Nauels A, Xia, Y, Bex V, Midgley PM. (Eds.), 2013. Summary for policymakers. In: climate change 2013: the physical sci- ence basis. In: Contribution of Working Group I to the Fifth Assessment Report of the Intergovernmental Panel on Climate Change. Cambridge University Press, Cambridge, United Kingdom and New York, NY, USA.

Venkateswarlu B and Shanker AK. 2009. Climate change and agriculture: Adaptation and mitigation strategies. *Indian J. Agronomy* 54: 226-230.

Venkateswarlu B and. Rao VUM. 2015. Agriculture in Madhya Pradesh. Climate change in Madhya Pradesh: A compendium of expert views-II, MPSKCCC, *Environmental Planning and Coordination Organization*, Bhopal

Wien HC, 1997. The Physiology of Vegetable Crops. Cab International, Oxon-New York.

WMO 2013. The Global Climate 2001-2010, A decade of climate extremes summary report. World Meteorological Organization, Geneva, Switzerland. Downloaded from http: //library.wmo.int/pmb_ged/wmo_1119_en.pdf

Chapter 11

Advances and Challenges in Eggplant Improvement

Pallavi Mishra[1] *and Shailesh K. Tiwari*[2]

[1]*Senior Research Fellow and* [2]*Senior Scientist*
Division of Crop Improvement,
ICAR-Indian Institute of Vegetable Research, Varanasi – 221 005, Uttar Pradesh
**Corresponding author*

Eggplant (*Solanum melongena* L.) commonly called aubergine or brinjal is an important vegetable crop known for therapeutic and culinary uses since ancient times. It also, is among the few Solanaceous crops belonging to the *'Old world'*, originating in home-country India and native sub-continents, where the greatest diversity among fruit characteristics is observed. Eggplant belongs to the division Magnoliophyta, class Magnoliopsida, subclass Asteridae and order Solanales in the Solanaceae family. This family comprises of economically important crops other than brinjal such as tomato, potato, pepper, tobacco, tomatillo, petunia, *etc.* Eggplant is grown worldwide for its edible spongy fruit which is used in cuisines and relished as dish. Although it is popular as a vegetable crop in India, various plant parts are also used as medicines for treatment of inflammations, bronchitis, asthma, cardio-vascular diseases and some forms of neuralgia. Cultivar differences in brinjal mainly concerns the fruit shape, size and weight, peel color, chemical composition, earliness, ability to withstand environmental-posed challenges and resistance to pests and pathogens hampering brinjal cultivation. Therefore, dissecting the genetic basis of variability among genotypes is potentially relevant for eggplant breeding programs. In the past few decades, remarkable progress has been achieved in conventional breeding methods through the introduction of advanced biotechnology-based breeding techniques and genome editing techniques, which has speed up the eggplant improvement program to a next level. Here, we summarize the general eggplant description, occurrence and diversity, nutraceutical features, advancements in research areas with special emphasize on biotechnology, and major challenges need to be addressed for all round genetic improvement of this vegetable crop.

Introduction

Solanum represents one of the most hypergenus group among the angiosperms comprising nearly 1,500 species. Many Solanum species are widely used as vegetables and also as popular medicines due to presence of steroidal alkaloids and phytochemicals of human benefit. Eggplant is an important vegetable crop of this genus which has been known for its culinary and therapeutic uses since the evolution of Ayurveda. The cultivar differences concern mainly the peel color, shape, size and weight of comestible fruits, but the chemical composition, earliness of fruiting, and environmental-posed challenges also play a significant role in genesis of variability and acquisition of different morphological forms. Morphological variation in eggplant is an important factor that governs the choice of fruit for human consumption. The genetic differences among the genotypes are potentially relevant to breeding programs in that the variability created through hybridization of the contrasting forms could be exploited. Although it is popular to be used a vegetable in India, the various plant parts are also used as medicines for treatment of inflammations, cardiac disabilities, bronchitis, asthma and several forms of neuralgias (Mutalik *et al.,* 2003).

Brief History of Eggplant

Eggplant (*Solanum melongena* L.) is grown worldwide for its edible fruit. As a member of the genus Solanum, it is related to tomato and potato. Most commonly seen forms are purple one which is the spongy absorbent fruit used in several cuisines. Eggplant and their progenitor species are believed to be originated in India (Arumugam and Anandakumar 2013; De Candolle 1885; Knapp *et al.,* 2013) but the first cultivation began in China in the 5th century B.C. which is now reported to be the secondary center of eggplant origin and domestication (Hurtado *et al.,* 2012; Lester and Hasan 1991). Therefore, it is also called to be a vegetable of 'Indo-Chinese' center of origin (Vavilov 1951). Eggplant is usually used as vegetable but the ancient history also describes the therapeutic uses of eggplant (Meyer *et al.,* 2014). Eggplant is usually a warm-weather crop but the related species are dispersed in different climatic zones from arid and semi-arid zone to desserts with extremely high temperature (Knapp *et al.,* 2013).

Origin of Eggplant

Eggplant and their progenitor species are known to be originated in India (Arumugam and Anandakumar 2013; Knapp *et al.,* 2013) but the first cultivation perhaps began in China in the 5th century B.C which is now reported to be the secondary center of eggplant origin and domestication (Hurtado *et al.,* 2012; Lester and Hasan 1991). It was then introduced to Africa in the late 14th century from when its cultivation subsequently spread to Europe and later to the other countries such as Turkey, Japan, and Italy (Knapp *et al.,* 2013). Ancient history also describes the therapeutic uses of eggplant (Meyer *et al.,* 2014). Europeans opted eggplant more to be used as a decorative plant rather than food plant as they believed the fruits were bitter in taste (Daunay *et al.*, 2007). Due to continuous domestication, the fruits lost its bitter taste and today, it is preferred as a part of cuisine all over in Asia and middle-east countries.

Taxonomical Hierarchy

Species level taxonomy of Solanum is always challenging due to presence of large number of species distributed around almost all the parts of the world in every variety of habitats. The genus is sub-divided into 13 major clades, the largest of which is the Leptostemonum clade comprising nearly 450 species including eggplant (Knapp *et al.*, 2013). Eggplant exists in three common cultivated forms, *S. melongena* (brinjal eggplant), *S. aethiopicum* (scarlet eggplant), and *S. macrocarpon* (gboma eggplant) which are native to Europe, Asian, and African countries. They are well-suited for phenotyping and agronomical studies as they exist in several morphological forms. The European Eggplant Genetic Resources Network (EGGNET; van der Weerden and Barendse (2006)) and the International Board for Plant Genetic Resources (IBPGR, 1990) have developed various morphological descriptors to characterize the useful species of eggplant for use in breeding programs.

Distribution of Eggplant and Species Diversity

Eggplant is an important crop of subtropics and tropics and grown extensively in India, Bangladesh, Pakistan, China and Philippines. It is believed to be originally domesticated from the wild nightshade species *S. incanum* also known as bitter apple, with two independent domestications centers, one in South Asia and another one in East Asia. Crop Wild relatives (CWR) of eggplant are good choice for interests in breeding as they render an extremely valuable gene pool to draw useful alleles required in breeding programs. Some wild forms used in breeding and are related to common eggplant includes *S. insanum*, *S. incanum*, *S. macrocarpon*, *S. aethiopicum*, *S. torvum*, *S. khasianum*, *S. sisymbrifolium*, *S. indicum*, *S. nigrum*, *S. xanthocarpum*, *S. gilo*, *etc.*

Improved Varieties

Cultivated eggplants exists in three most common forms, *S. melongana var. esculentum* which is the common aubergine including many cultivars, *S. melongena var. depressum* which is dwarf aubergine, and *S. melongena* var. serpentium which are the snake aubergines. Many other forms of eggplant types, such as *agreste, album, divaricatum, esculentum, giganteum, globosi, inerme, insanum, leucoum, luteum, multifidum, oblongo-cylindricum, ovigera, racemiflorum, racemosum, ruber, rumphii, sinuatorepandum, stenoleucum, subrepandum, tongdongense, variegatum, violaceum, and viride* are also known which are considered as different cultivar groups of eggplant. One break-through in eggplant improvement has been mediated through genetic engineering; the *Bt*-brinjal developed by inserting the *cry1Ac*-gene from *Bacillus thuringiensis*. *Bt*-brinjal was developed with a view to have complete control over brinjal fruit and shoot borer (BFSB) without affecting non-target arthropod biodiversity, reduced insecticide sprays and up to six fold increase in the farmer's profit. But unfortunately, it sought a ban from the Parliamentary Committee on Agriculture in 2012 in view of nutritional security and toxicity-assessment upon human health. Presently, field trails for GM brinjal are under practice at the ICAR-IARI, New Delhi; ICAR-IIHR, Bengaluru; ICAR-IIVR, Varanasi, and NIPGR, New Delhi.

Varietal differences among Various Forms of Cultivated Wild and Eggplants

Crop wild relatives (CWRs) or in general, the wild relatives of agricultural crops exhibit enormous diversity compared to the domesticated form and possess the capacity to develop more productive, disease resistant, and resilient varieties. The wild relatives offer an extremely valuable gene pool to draw useful alleles required in breeding programs. For the crops with narrow genetic base, the wild relatives act as a valuable source for transfer of genes of desired traits of interest to be used in breeding and broadening the genome of lesser diverse lines (Dempewolf *et al.*, 2017; Zhang *et al.*, 2017). CWRs of eggplant are richly distributed along the globe towards Africa, Asia, India, and eastwards to Pakistan, and these regions are supposed to be the center of eggplant origin and domestication. Farming practices, breeding efforts, changing climatic factors, and increasing pressures of the market demands lead to subsequent alteration in the morphological and genetic individuality of CWR of eggplants (Dempewolf *et al.*, 2017). The primitive plant and fruit characters include broad, spiny foliage, small and large number of fruits, bitter taste, tough peel and thick flesh. Systematic efforts are made by breeders for improvement of the nutritive value, flavor, aroma, and yield of fruits with better quality and soft flesh with fewer seeds.

Nutritional Value and Phyto-chemistry of Brinjal

100 gram of a raw eggplant fruit comprises 92 per cent water, 6 per cent carbohydrates, 1 per cent protein, and has negligible fat. Minor changes in nutrient composition may occur with changes in climate or environment of cultivation and genotype. Although it provides low amounts of essential nutrients with only manganese having a moderate percentage (11 per cent) of the daily value, fruits are rich source of dietary fibers, Vitamin B_1, B_3, B_6, C, K, pantothenic acid, beta-carotene equivalents and folate. Fruits also contain arginine, aspartic acid, histidine, delphinidine–3-bioside (nasunin), oxalic acid, solasodine, ascorbic acid, tryptophan, *etc.* Leaves contain rich amounts of chlorogenic, hydrocaffeic and protocatechuric acids. Some of the alkaloids present are tropane, pyrrolidine, quinazolizidine, steroid alkaloids and glycoalkaloids. Two steroidal saponinsmelongoside-L and melongoside-M, and three new saponins melongoside-N, -O and P, have been isolated from eggplant seeds. Catechol oxidase has been isolated and characterized from *S. melongena*. Another bioflavonoid glycoside, *solanoflavone* present in the leaves and fruits of *S. melongena* has also been isolated.

Medicinal and Culinary Uses of Eggplant

The medicinal properties of the plant are derived from its chemical constituents (Das and Barua 2013). The plant's antioxidant property is due to the presence of flavonoids (Akanitapichat *et al.*, 2010). The presence of terpenes makes it useful for bronchitis (Yang *et al.*, 2020). Analgesic property is because of the alkaloids. Eggplant is also known to possess anti-pyretic, antioxidant, anti-asthamatic and spasmogenic activities, and hypotensive, anaphylactic and hypo-lipidemic actions. Besides these, the biological activities of alkaloids such as *calystegines* and *nortropane* isolated from eggplant have also been described in regulating normal heart and liver functioning (Andersson 2002; Asano *et al.*, 1997). Eggplant contains high amount of dietary fiber which benefits to manage blood cholesterol levels. Chlorogenic acid (CGA) which is a primary antioxidant in eggplants are known to decrease the levels of low density lipoprotein (LDL) or bad cholesterol and reduces the risk of fatty liver diseases (Koley *et al.*, 2019). The polyphenols present in eggplant helps protect the body from severe forms of cancer (Kaushik 2019). Anthocyanins and chlorogenic acid are also known to protect tissues from damage caused by free radicals (Plazas *et al.*, 2013). In the long term, this may help prevent tumor growth and the spread of cancer cells. Anthocyanins also helps achieve this by preventing new blood vessels from forming tumor, reducing inflammation, and blocking the enzymes that help cancer cells spread (Liu *et al.*, 2020). Other research on animal studies has suggested that nasunin, an anthocyanin in eggplant skin helps protect the brain cell membranes from damage caused by free radicals. Nasunin also helps transport nutrients into cells and move waste out.

Common Diseases of Cultivated Eggplant

Being a warm-season crop with longer growth period, eggplant is exposed to wide range of pests, nematodes, and pathogens hampering its yield and productivity. Eggplant productivity is diminished by common pests like aphids, whiteflies, beetles, red spider mite, leafhopper, leaf roller, stem borer and fruit and shoot borer, and also by various soil-borne pathogens of bacterial and fungal origin causing common eggplant diseases like bacterial wilt, fusarium wilt, verticillium wilt, damping-off, phomopsis blight, phytophthora blight, fruit rot, leaf spot, and little leaf of brinjal (Magioli and Mansur 2005). Due to lack of resistance in the genome of cultivated eggplant, and insufficient characterization of wild relatives offering potential candidate resistance genes, breeding strategies are often limited in the crop. Therefore, there is a need to improve farming practices and develop more resistance eggplant varieties through utilization of conventional breeding strategies blended with modern biotechnological techniques to overcome these challenges. Eggplant breeding program for disease resistance mainly focuses on utilization of F_1 hybrids, as they exhibit high morphological variability, good shelf-life, and better quality fruits with well-developed internal resistance mechanisms (Rotino *et al.*, 1997). Some of the most vigorous pathogens and pests hindering eggplant productivity and yield are listed below in **Table 11.1.**

Table 11.1: Common Diseases of Eggplant Caused by Pests and Pathogens

Diseases	*Causative Agents*	*Symptoms*
Damping-off	Species of *Pythium, Fusarium, Phytophthora,* and *Rhizoctonia*	Seedlings fail to develop, rotting of taproot resulting into inability to absorb water, plants dry up and collapse
Phomopsis fruit rot	*Phomopsis vexans*	Circular pale to brown lesions over fruit with sunken, lighter areas which may cause depressions and softening of fruit
Cercospora leaf spot	*Cercospora melongenae*	Small circular yellow spots forming concentric dark brown colored rings, causing leaf chlorosis and finally necrosis
Alternaria leaf spots	*Altrenaria melongenae*	Leaf margins become yellow to dark brown which leads to necrosis, infection spreads to entire plant, necrotic spots and brown lesions over fruit, fruit hardening
Powdery mildew	*Leveillula taurica*	White powdery spots over shoot, flowers, and fruits, twisted yellow colored leaves which finally drop from the plants
Verticillium wilt	*Verticilium dahliae*	Yellow patches over lower surface of leaves, lead edges become thin and curl inwards, symptoms spread to entire plant, complete blockage of vascular system resulting into stunted growth and wilting
Bacterial wilt	*Pseudomonas solana-cearum, Ralstonia solanacearum*	Taproot is most affected by this soil-borne pathogen, slow plant growth, browning of vascular tissues which affects underlying tissues, yellowing of leaves and permanent wilting occurs
Little leaf of brinjal	Phytoplasma, transmitted by *Hishimonus phycitis*	Reduced lead size and rosette-like appearance, affected plant cannot bear fruit or have hard inedible fruits
Eggplant fruit and shoot borer	*Leucinodes orbonalis*	Most destructive of all kind of eggplant diseases and pests, pest is active in almost all seasons and can infest both the vegetative and fruiting stage, larvae feed the fruits and cause rotting, limited plant growth and yield

Breeding Status in Eggplant

Conventional breeding efforts in eggplant were focused particularly in the massive production of cultivated variety to obtain good quality and resistance varieties producing large number of fruits for therapeutic and culinary reasons. Conventional breeding strategies has been reviewed time to time to provide an insight into progress achieved in eggplant breeding programs (Kalloo and Bergh, 2012). Several efforts for breeding for quantitative and qualitative traits have resulted in introduction of improved eggplant varieties into domestication. Since the cultivated variety is vulnerable to wide range of biotic and abiotic stresses, one of the main problems of traditional breeding is to insert the resistance genes for improved eggplant yield and fruit quality traits. Therefore, resistance to pathogens and pests has also been identified in many wild species, mostly belonging to sub-genus *Leptostemomocus* of family Solanaceae, which are further used as breeding materials for introgression of suitable alleles to cultivated variety (Daunay and Lester, 1988). Many interspecific hybrids have been successfully developed so far using *S. melongena* and wild species such as *S. indicum, S. macrocarpon, S. khasianum, S. aethiopicum, S. insanum, S. gilo, S. torvum, S. integrifolium, S. sisymbriifolium, S. xanthocarpum etc.* (Daunay and Lester 1988; McCammon and Honma 1983;

Sharma 1980; Sharma *et al.*, 1984). Eggplant genotypes are consistently deployed in developmental, cytological and biotechnological research for development of improved varieties. Due to the high morphogenetic potential of its tissues, eggplant species can be easily subjected to cytological as well as biotechnological studies such as tissue culture, organogenesis, haploidisation, embryo rescue, genetic transformations, somatic hybridization, karyotype analysis and *in vitro* regeneration to overcome the problems of sexual incompatibilities appearing during the cross of cultivated variety with different wild ancestors, and have been reviewed time to time by Kashyap *et al.* (2003), Saini and Kaushik (2019), and Sreekar (2020). Several studies elaborating the use of eggplant and their wild relative to exploit heterosis breeding strategies for increased productivity, development of eggplant cultivars with enhanced quality and yield, and development of hybrids with multiple resistances to insects and pests have been reported till date (Fassuliotis *et al.*, 1981; Kashyap *et al.*, 2003; Kirti *et al.*, 1984; Rao and Baksh 1979; Venora *et al.*, 1992). However, one cannot always depend to develop the new varieties with desirable traits using conventional breeding strategies only, as it is mostly dependent on the phenotypic observation and selection methods. To get much deeper insights, one need to open the genome which could be achieved through adopting molecular marker based selection, marker assisted breeding and other emerging techniques such as biotechnology, genome engineering, molecular farming and use of '*omics*' technology as discussed later. In general, improvement of yield and fruit quality of eggplant, while making it resistant to adverse scenarios caused due to biotic/abiotic stresses by inserting resistance genes through the combined use of conventional breeding and biotechnological approach, are the main areas of focus in eggplant breeding.

Marker Assisted Breeding

Characterization of genetic diversity is fundamental to follow an efficient breeding program particularly for crops like eggplant exhibiting high level of heterogeneity. Molecular marker-assisted breeding methodology is a revolution in crop improvement programs as has well-support the conventional breeding strategies, since the traditional phenotypic assays are not sufficient to assess a particular trait. The cytological studies carried out earlier can be well supported and presented with proper evidence obtained through DNA fingerprinting technique. For this, short stretches of arbitrary sequences called molecular markers are used which are representatives of a particular DNA segment and can directly be correlated with a phenotype (Collard *et al.*, 2005). The cost efficacy, stability, reliability and high reproducibility of these markers permits the usage of these markers for construction of genetic linkage maps, QTL mapping, MAS, assessment of genetic diversity within genotypes and gene pyramiding (Stuber *et al.*, 1999; Varshney *et al.*, 2007; Wu *et al.*, 2009b) and provides quick solutions to genetic problems like undesirable wild features and unfavorable scenarios like climate and environmental changes (Thapa *et al.*, 2015). In case of eggplant, molecular diversity have been studied using array of DNA-based markers, like randomly amplified polymorphic DNAs (RAPDs), amplified fragment length polymorphisms (AFLPs), restriction fragment length polymorphisms (RFLPs), inter-simple sequence repeats (ISSRs), simple sequence repeats (SSRs) including sequence-tagged microsatellite sites (STMS), single nucleotide polymorphism (SNPs) *etc.*, which have been extremely useful for germplasm characterization (Baysal *et al.*, 2010; Furini and Wunder 2004; Isshiki *et al.*, 2008; Isshiki *et al.*, 1994; Isshiki *et al.*, 2003; Isshiki *et al.*, 1998; Karihaloo *et al.*, 1995; Karihaloo and Gottlieb 1995; Nunome *et al.*, 2009; Nunome *et al.*, 2003; Tiwari 2007; Tiwari *et al.*, 2009), assessment of genetic diversity (Ansari *et al.*, 2015; Behera *et al.*, 2006; Karihaloo *et al.*, 2002; Koundal *et al.*, 2006; Sakata and Lester 1994; Singh *et al.*, 2006), establishment of taxonomy-based genetic relationships and phylogeny among diverse eggplant accessions (Isshiki *et al.*, 1998; Lester and Hasan 1991; Sakata *et al.*, 1991) and QTL mapping studies (Barchi *et al.*, 2012; Doĝanlar *et al.*, 2014; Doganlar *et al.*, 2002a; Frary *et al.*, 2014; Fukuoka *et al.*, 2012; Gramazio *et al.*, 2014; Lebeau *et al.*, 2013; Miyatake *et al.*, 2012; Nunome *et al.*, 2001; Nunome *et al.*, 2009; Nunome *et al.*, 2003; Salgon *et al.*, 2017; Sunseri *et al.*, ; Wu *et al.*, 2009a). Some gene-targeted molecular markers like Start Codon Targeted markers (SCoT) are a good alternative to replace the conventional molecular marker system and have been reported in many other crops peanut (Xiong *et al.*, 2011), duram wheat (Etminan *et al.*, 2016), citrus (Mahjbi *et al.*, 2015), ramie (Satya *et al.*, 2015), rice (Collard and Mackill 2009), dendrobium (Bhattacharyya *et al.*, 2013), jatropha (Mulpuri *et al.*, 2013), mango (Luo *et al.*, 2012) and tomato (Shahlaei *et al.*, 2014) for assessment of diversity and genetic relationships among different cultivars, but not yet on eggplant. Overall, the studies from previous reports show that molecular marker technique is a worthy tool for identification of key genotypes with contrasting traits or alleles in population (Ansari *et al.*, 2015) and selection of the genotypes with desirable traits of interest (Agarwal *et al.*, 2008; Kardavan 2015), thereby enhancing the efficiency of marker assisted selection.

Tagging of Genes/QTLs Controlling Traits of Agronomic Interest in Eggplant

Quantitative genomics is a well-defined area of study in the family Solanaceae. Quantitative genomics assumes that relatively small number of loci exits in the genome of plants which controls their physiological and morphological characteristics, and the key source of natural variation existing between the populations (Kearsey and Farquhar 1998; Paran and Zamir 2003). Identification of these loci allows the farmers and breeder to distinguish the cultivated crops varieties from their wild relatives based on domestication traits (Doganlar *et al.*, 2002b; Paterson 2002). These loci are also called polygenes or quantitative traits loci (or QTLs), and are involved in processes such as gene-to-gene vs. gene-to-environment interactions, gene activation and gene transmission among the population (El-Soda *et al.*, 2014). QTLs are related to a particular genotype and directly affect the phenotype of an individual (Paran and Zamir 2003). Precisely locating the position of QTL with the help of molecular markers (for example, RAPD, ISSR, SSR, SCoT, DArT, SCAR *etc.*) is a great achievement in the in the area of quantitative genomics, and is referred to as QTL mapping. QTL mapping is an efficient method to locate the position of genes controlling plants physiological, biochemical, molecular and morphological characteristics by unraveling the complex traits (Kearsey and

Farquhar 1998). QTL analysis predicts the association between marker alleles segregating in an experimental population and the quantitative characters inherited by the population (Collard *et al.*, 2005). The first interspecific genetic linkage map in eggplant was reported to locate the QTLs controlling fruit shape in eggplant using RAPD markers by Nunome *et al.* (1998). Several linkage and QTL mapping studies have been extensively carried out in eggplants to dissect the genetic basis of variation underlying horticultural, resistance and yield-related traits of economic importance in eggplant (Barchi *et al.*, 2012; Docimo *et al.*, 2016; Doganlar *et al.*, 2002b; Frary *et al.*, 2003; Frary *et al.*, 2014; Gramazio *et al.*, 2014; Kaushik 2019; Lebeau *et al.*, 2013; Mishra *et al.*, 2020; Miyatake *et al.*, 2012; Miyatake *et al.*, 2016; Miyatake *et al.*, 2020; Mutlu *et al.*, 2008; Nunome *et al.*, 2001; Portis *et al.*, 2014; Rinaldi *et al.*, 2016; Toppino *et al.*, 2016; Wei *et al.*, 2020b), as listed in **Table 11.2**.

Biotechnological Tools for Improvement of Eggplant Genetic Resources

Tissue Culture Techniques

Eggplant presents high morphogenetic potential of it tissues for *in vitro* techniques such as tissue culture (Kamat and Rao 1978), which allows its uses in broad areas of application in biotechnology like cytological and karyotype studies, embryo rescue, somaclonal variation, somatic hybridization, haploidisation, and gene transfer using *Agrobacterium*-mediated genetic transformation. *In vitro* organogenesis of eggplant can be easily achieved through addition of balanced amount of macro-and micro-nutrients and plant growth regulators (PGRs), and was first reported in *S. sisymbriifolium*, a wild species of eggplant (Fassuliotis *et al.*, 1975). This was achieved by isolating and culturing shoot parenchyma cells in Linsmeier–Skoog (LS) medium supplemented with 6-(γ,γ-dimethylallylamino)-purine (2iP) and indole-3-acetic acid (IAA). Shoot regeneration has been successfully carried out many wild species of eggplant such as *S. indicum*, *S. xanthocarpum*, *S. aethiopicum*, *S. khasianum*, *S.torvum*, *S. aviculare etc.* (Fassuliotis *et al.*, 1981; Gleddiei *et al.*, 1985; Guri *et al.*, 1987; Kowalczyk *et al.*, 1983; Rao and Narayanaswami 1968).

Plant regeneration using protoplasts is potentially useful in overcoming pre- and post-fertilization breeding barriers encountered during conventional breeding, and has been reported in cultivated and wild species of eggplant (Fournier *et al.*, 1995; Sihachakr and Ducreux 1987). Protoplast fusion technique is a milestone in this regard and has facilitated the transfer of useful traits among sexually incompatible eggplant species through the process called somatic hybridization. Successful somatic hybrids have been developed using *S. melongena* and wild species of eggplant, and also with other species like tomato and tobacco (Collonnier *et al.*, 2001; Daunay *et al.*, 1993; Gleddie *et al.*, 1986; Guri *et al.*, 1991; Jarl *et al.*, 1999; Liu *et al.*, 1995; Toki *et al.*, 1990). Somatic hybridization is very efficient method to introduce useful traits from wild species like resistance to bacterial and fungal wilts, and nematodes resistance into cultivated variety. For instance, successful somatic hybrids developed using *S. melongena* and *S. aethiopicum* are shown to be highly resistant to bacterial wilt disease caused by *Ralstonia solanacearum* (Collonnier *et al.*, 2001).

Table 11.2: QTL Mapping Studies Reported in Eggplant and Related Species

Eggplant Genotypes	*Population*	*Traits*	*Markers*	*Linkage Groups and Map Length*	*Average Marker Density*	*Reference*
EPL-1(*S. melongena*) x WCGR112-8 (*S. melongena*)	168 Intraspecific F_2	Fruit shape and color	93 AFLPs,88 RAPDs	21,779 cM	4.9 cM	Nunome *et al.* (2001)
MM195 (*S. linnaeanum*) xMM738 (*S. melongena*)	58 Interspecific F_2	Fruit domestication traits	232 RFLPs	12,1,480 cM	7.6 cM	Doganlar *et al.* (2002b)
EPL-1 (*S. melongena*) xWCGR112-8 (*S. melongena*)	120 Intraspecific F_2	Fruit shape and color development	101 RAPDs54 AFLPs7 SSRs	17,716 cM	4.9 cM	Nunome *et al.* (2003)
EPL-1 (*S. melongena*) xWCGR112-8 (*S. melongena*)	94 Intraspecific F_2	SSR development	245 SSRs	14,959 cM	4.3 cM	Nunome *et al.*, 2009
305E40 (DH from *S.melongena* and *S.aethiopicum*) x 67/3 (*S.melongena*)	156 Intraspecific F_2	Anthocyanin content of fruit	339 SNPs33 SSRs27 COSII11 RFLPs3 CAPS2 HRM	12,1,390 cM	3.8 cM	Barchi *et al.* (2012)
LS1934 (*S.melongena*) xWCGR112-8 (*S.melongena*)LS1934(*S.melongena*) x AE-P03 (*S.melongena*)	90 F_2 Intraspecific (LWF2) 93 F_2 Intraspecific (ALF2)	Development of Solanum orthologous (SOL) sets	639 SNPs313 SSRs	12,1,285 cM	1.4 cM	Fukuoka *et al.*, 2012
LS1934 (*S. melongena*) xAE-P03 (*S. melongena*)Nakate- Shinkuro (*S. melongena*) x AE-P03 (*S. melongena*)	135 Intraspecific F_2 (ALF2) 93 Intraspecific F_2 (NAF2)	QTLs associated with parthenocarpy	132 SNPs118 SSRs125 SSRs49 SNPs	12,1,414 cM15,1,153 cM	--	Miyatake *et al.*, 2012
MM738 (*S. melongena*) xMM960 (*S. melongena*)	178 Intraspecific F_6 RILs	Genetic control of resistance to *R. solanacearum*	91 AFLPs26 SSRs2 SRAPs	18,884 cM	8.8 cM	Lebeau *et al.*, 2013
Eggplant accessions fromUSA, India, Japan, Italy, Malaysia, Arab Emirates, Thailand, Korea	141 Association mapping (GWA)	Association analysis for fruit morphological traits	105 SSRs	-	-	Ge *et al.* (2013)
MM195 (*S. linnaeanum*) x MM738(*S. melongena*)	108 Interspecific F_2	Synteny with tomato	400 AFLPs348 RFLP116 COSII	12,1,518 cM	1.8 cM	Doganlar *et al.*, 2014

Eggplant Genotypes	*Population*	*Traits*	*Markers*	*Linkage Groups and Map Length*	*Average Marker Density*	*Reference*
MM577(*S. incanum*) x AN-S-26(*S. melongena*)	91 Interspecific BC1	Mapping CGA biosynthesis pathway and genes for polyphenoloxidase (PPO)	99 SSRs88 AFLPs42 COSII9 CAPS4 SNPs, and 1 morphological marker	12,1,085 cM	4.4 cM	Gramazio *et al.*, 2014
Eggplant breeding lines,old varieties,and landracesfrom Asia andMediterranean	191, Association mapping (GWA)	Fruit, plant, and leaf morphologica ltraits	314 SNPs	-	-	Portis *et al.* (2015)
MM738 (*S.melongena*) xMM960 (*S.melongena*)	180 Intraspecific F_2	Genetic control of resistance to *R. solanacearum*	867 SNPs139 AFLPs28 SSRs1 SRAP	14,1,518 cM	1.4 cM	Salgon *et al.*, 2017
S. melongena x*S. incanum*	114 interspecific RILsF_8	Morphological and yield-related traits	379 RAPD,944 SSRs,40 ISSR,80 SCoT	12,1165 cM	4.13 cM	(Mishra *et al.*, 2020)

Somatic embryogenesis or the process of development of an explant or embryo using single somatic cell, has been reported in eggplant using stems, leaves, roots, cell suspension, hypocotyl and protoplasts (Gleddie *et al.*, 1983; Matsuoka and Hinata 1979; Singh Yadav and Venkat Rajam 1998). First study of regeneration from anther culture in eggplant was reported by Raina and Iyer (1973). Anther culture has been successfully used to reduce the ploidy level of somatic hybrids developed using *S. melongena* and wild species such *S. aethiopicum* gilo group (Rizza *et al.*, 2002).

Genetic Engineering

The first successful genetic engineering in eggplant was reported in 1980s using *Agrobacterium*-mediated genetic transformation with a co-integrate vector harbouring *nptII* and *hptII* as marker genes (Guri and Sink 1988). This opened new gateways for genetic transformation techniques in eggplant using different gene constructs, selectable markers (for *e.g. nptII*) and reporter genes like β-glucouronidases (*GUS*), chloramphenicol acetyl transferases (*CAT*) and luciferase. Several gene transformation protocols have been standardized in eggplants for development of varieties with improved resistance to insects like fruit and shoot borer and beetles (Arpaia *et al.*, 1997; Gojko *et al.*, 1998; Hamilton *et al.*, 1997; Iannacone *et al.*, 1997; Kumar *et al.*, 1998), abiotic stresses (Donzella *et al.*, 2000; Prabhavathi *et al.*, 2002), antibiotics (Sharon *et al.*, 1997), induction of parthenocarpy (Donzella *et al.*, 2000; Rotino *et al.*, 1997), and accumulation of Bcarotene in eggplant fruits (Mishiba *et al.*, 2020), as listed in **Table 11.3**.

Table 11.3: Genetic Engineering for Development of Transgenic Eggplant

Marker Gene	*Transformed Gene*	*Purpose*	*Reference*
nptII and *hptII*	-	Successful *Agrobacterium*-mediated genetic transformation	Guri and Sink, 1988a, Guri and Sink, 1988b
nptII	*Bt*(*CryIIIB*)	Introduction of modified Bt-gene into eggplant for resistance to Colorado Potato Beetle (CPB). The transgenic eggplant were highly resistant to CPB	Arpaia *et al.* (1997)
nptII	Modified *Bt*(*CryIIIA*)	Development of transgenic eggplants expressing modified Bt-gene	Hamilton *et al.* (1997)
nptII	*iaaM–DefH9*	Introduction of gene inducing parthenocarpy in eggplant. As a result, seedless eggplant fruits were developed	Rotino *et al.* (1997)
nptII	Modified *Bt*(*CryIII*)	Introduction of modified *Bt* (*CryIII*) gene confirmed resistance in cultivated eggplant and wild species *S. integrifolium*	Iannacone *et al.* (1997)
nptII	Synthetic *Bt*(*CryIAb*)	Introduction of synthetic Bt-gene for expression of insecticidal protein, transgenic plants thus produced exhibited protection against fruit and shoot borer (*Leucinodes orbonalis*)	Kumar *et al.* (1998)

Marker Gene	Transformed Gene	Purpose	Reference
nptII	Synthetic *Bt(CryIIIA)*	Development of transgenic eggplant resistant to neonate larvae and adult CPB. The segregating F_1 population also exhibited resistance to CPB	Jelenkovic *et al.* (1998)
nptII	*DefH9–iaaM*	Winter cultivation of eggplant in an unheated greenhouse resulted in higher winter production and improved fruit quality	Donzella *et al.* (2000)
nptII	*mtlD*	Introduction of mannitol-1-phospho-dehydrogenase (*mtlD*) gene conferring tolerance to salinity and drought induced osmotic stress in eggplant	Prabhavati *et al.* (2002)

Molecular Farming and Prospects for Eggplant

With the advancements in genetic engineering technique, it is now possible to engineer genes of interest to produce therapeutically active substances in simple yeasts to the more complex plants. This method is simply called Bio-farming, and leads to production of biopharmaceuticals. Molecular farming is an emerging technology (technically the method similar as used to produce genetically modified or GM crops), that uses different organisms like yeasts, fungi, bacteria and even plants for production of pharmaceutical substances such as vaccines and antibodies in extremely large quantities (Fischer and Emans 2000; Ma *et al.*, 2003; Obembe *et al.*, 2011). Plant bio-farming is far more advantageous than the use of yeast, fungi or bacteria as models, as the huge quantities of recombinant pharmaceutical and industrial proteins can be produced in plants with low cost and input requirements and no risks or toxicity upon human health and environment (Bhatia and Dahiya 2015; Goldstein and Thomas 2004). Due to the abundance of polyphenolics, anti-oxidants, vitamins and many phyto-active compounds in wild and cultivated eggplant species, these are of potential uses to pharmaceutical industries for effective drug delivery. Eggplant cultivars could be engineered for higher production and accumulation of antioxidants for increased culinary uses. Eggplant generally contains low protein content, therefore through molecular farming approach; molecular mechanisms controlling the pathways for protein synthesis can be triggered for higher protein production and accumulation in eggplant. In the recent years, variety of plants such as Arabidopsis, carrot, tomato, capsicum, cauliflower, cucumber and cranberries including eggplant have been used as model plants for large-scale production of edible vaccines, plant-derived antibodies (or plantibodies), industrial enzymes and recombinant proteins (Bhatia and Dahiya 2015; Goldstein and Thomas 2004; Rajput *et al.*, 2007; Singh *et al.*, 2010).

Genome Engineering

Genome engineering is an emerging technique which uses a variety of modification techniques to make it possible to modify, remove, replace or insert unique genomic sequences of interest into a targeted site in the genome of host organism (Gaj *et al.*, 2013; Ran *et al.*, 2013). It is quite different from genetic transformation method in which a gene of interest is randomly inserted into the host genome at site-specific locations following the genome editing guides (Carr and Church 2009). With the emergence of highly specific and cost-efficient genome editing techniques and availability of sequenced genome, scientists can now easily alter plant genomes directly at the desired location using sequence-specific endonucleases (Gaj *et al.*, 2013). In the recent years, a number of genome editing techniques such as Meganucleases, zinc-finger nucleases (ZFNs), transcription activator–like effector nucleases (TALENs) and Clustered Regularly Interspaced Short Palindromic Repeats (or CRISPR/*Cas9*) have emerged making it feasible to facilitate the editing of genomic loci more precisely (Ran *et al.*, 2013). Genome engineering using Meganucleases involves the use of sequence-specific endonucleases which recognizes very large sequence sites (>12 bp) and hence, are less specific in their mode of action (Galetto *et al.*, 2009). However, these are an efficient tool for mammalian gene therapy for treatment of diseases like Xeroderma pigmentosum, production of recombinant proteins, complementing virus-borne diseases *etc.* (Galetto *et al.*, 2009; Pâques and Duchateau 2007; Redondo *et al.*, 2008; Silva *et al.*, 2011). In contrast, the other techniques (ZFNs, TALENs, CRISPR/*Cas9*) are more specific and accurate, and can precisely edit the target DNA site of host genome (Hsu *et al.*, 2014). CRISPR/*Cas9* system is widely used in genome engineering in variety of plants including popular Solanaceous vegetables such as potato, tomato and eggplant (Andersson *et al.*, 2018; de Toledo Thomazella *et al.*, 2016; Enciso-Rodriguez *et al.*, 2019; Gianoglio *et al.*, 2018; Maioli *et al.*, 2020; Ueta *et al.*, 2017; Van Eck 2018; Wang *et al.*, 2015b; Zhi *et al.*, 2020).

Applications of '*Omics*' Technology in Eggplant Breeding

'*Omics*' is a new terminology which collectively combines the studies comprising genomics, transcriptomics, proteomics and metabolomics of either individual or pool of a biological sample. *Omics* study allows to look-up at the state of a sample at a particular time, and to identify the structure, functions and dynamics of a living organism. *Omics* tools have revolutionized the agricultural research by supporting the conventional plant breeding methodologies with modern tools and techniques that assists in enhancing the nutritional composition, quality, shelf-life and taste of vegetable crops, development of superior varieties with resistance to insects/pests and various kind of abiotic stresses, and tolerance to pesticides or herbicides. In other words, *omics* study integrated with analytical methods and computational analysis provides deeper insights into the molecular mechanisms underlying the traits of economic importance by simplifying the complex gene interactions, leading to crop improvement. Many improved vegetable varieties have been developed through the use of *omics* approach. In case of eggplant, this technique has greatly assisted the eggplant breeding program by large-scale development of molecular markers for fingerprinting and diversity analysis (Barchi *et al.*, 2011; Gramazio *et al.*, 2016; Gramazio *et al.*, 2017; Portis *et al.*, 2018), gene mapping and QTL tagging (Mishra *et al.*, 2020; Toppino *et al.*, 2020; Wei *et al.*, 2020b), domestication trends and phylogenomics studies (Wei *et al.*, 2019; Yang *et al.*, 2014), development of multiple resistance to diseases and abiotic stress (Li *et al.*, 2019; Ramesh *et al.*, 2016; Yang *et al.*,

2014; Yang *et al.*, 2019; Zhang *et al.*, 2020), triggering metabolic pathways affecting physiological and biochemical attributes (Li *et al.*, 2018; Li *et al.*, 2017; Wang *et al.*, 2017; Yang *et al.*, 2018; Zhang *et al.*, 2019), fruit quality traits (Chen *et al.*, 2017), and by providing draft and whole genome assembly of eggplant for improved future breeding practices (Gramazio *et al.*, 2019; Hirakawa *et al.*, 2014; Wei *et al.*, 2020a). Few studies characterizing *omics* tools in eggplant improvement are listed in **Table 11.4.**

Table 11.4: *Omics* Charactarization in Eggplant and Related Wild Species

Species	*Assembly Type and Plant Tissue*	*Sequencing Platform*	*Assembly Statistics*	*NCBI Accession*	*Reference*
S. melongena	Identification of MiRNA,Seedlings	Illumina MiSeq	5,940 miRNA	SRR833801 SRR833802	Yang *et al.*, 2013
S. melongena *S. torvum*	Transcriptome Assembly, Leaves, root, and stem	Illumina HiSeq 2000 PE (72 bp)	38,185 unigenes 34,174 unigenes	SRR1104129 SRR1104128	Yang *et al.*, 2014
S. melongena	Draft Genome Assembly, Leaves, roots, fruits, and flowers	Roche 454 GS FLX PE (200–300 bp)	33,873 scaffolds	DRR014074 DRR014075	Hirakawa *et al.*, 2014
S. melongena	Genome Assembly, Leaves	Illumina	12 pseudo-molecules	-	Barchi *et al.*, 2016
S. melongena	Identification of Putative allergens, Whole fruit	Illumina HiSeq 2000 PE (100 bp)	149,224 transcripts	SRR1291243	Ramesh *et al.*, 2016
S. aculeatis-simum	Transcriptome Assembly, Root	Illumina HiSeq 2000	69,824 unigenes	SRS1383901 SRS1383902	Zhou *et al.*, 2016
S. aethiopicum *S. incanum*	Transcriptome Assembly, Leaf, floral bud, and fruit	Illumina HiSeq 2000 PE (300 bp)	87,084 unigenes 83,905 unigenes	SRR2229192 SRR2289250	Gramazio *et al.*, 2016
S. melongena	Identification of MiRNA, Pistil	Small RNA library	686 miRNAs	SRR3479276 SRR3479277	Wang, 2017
S. melongena	Transcriptome Assembly, Flower buds	Illumina Hiseq 2000	1592 differentially expressed genes related to parthenocarpy	SRP085349	Chen *et al.*, 2017
S. melongena	Transcriptome Assembly, Eggplant peel	Illumina HiSeq 2000	32,629 transcripts 1956 differentially expressed genes	SRR5650714 SRR5651526 SRR5658205 SRR5658226	Li *et al.*, 2018

Conclusion and Prospective

Eggplant is an important vegetable fruit crop in the family Solanaceae. Due to several independent centers known for eggplant origin, crop shows enormous diversity because of which the species-level taxonomy of such large groups is always challenging. In eggplants, cultivar differences concern mainly the color, shape, and height of fruits, but the chemical composition of fruits, the earliness of fruiting, yield, environmental requirements, *etc.* also play a significant role in variability. The close relationship of many wild species with the cultivated type is a debatable subject because of complex species-level distinction at the molecular level which complicates the differentiation of the taxa. What brings novelties in a plant to acquire unique forms? Whether external factors such as environment and pollen dispersal play role in adaptive changes in morphology or is there any evolutionary genetic basis that facilitates the acquisition of a particular plant form? Studies suggest that the pre-existing gene forms acquire some subsequent modifications in response to interaction with climate and environmental factors resulting into changes in their alleles which further alters regulatory networks and gene expressions and imparts sudden morphological variations within the species (Wang *et al.*, 2015a). Moreover, association and interaction of several environmental, cellular, and genetic elements cause large morphological variance and development of unique fruit features with nutraceutical and pharmaceutical benefits (Gürbüz *et al.*, 2018). Fruit morphological variation is a complex trait especially for the species like eggplant which have under-exploited genetic resources and their utilization. An elaborated study rendering insights into developmental changes underlying morphological variability and enhancing the translational work in non-model species is of great future highlight to develop the strategies for improvement of the overall nutritional quality and productivity of fruits.

With the advancements of biotechnological techniques and tools, future of eggplant improvement seemingly appears to be brighter than the past years. Biotechnology has played a significant role in the improvement of this vegetable crop through facilitating the conservation of eggplant genetic resources, marker-assisted phenotype assessment and distinction of sub-species, haploidisation, overcoming sexual incompatibility barriers though somatic hybridization and cybridization techniques, *Agrobacterium*-mediated gene transfers for development of cultivars with better quality, and herbicide and insect resistance, induction of parthenocarpy *etc.* Molecular marker studies like AFLP, RAPD, RFLP, ISSR, SSRs and SNPs are exploited for eggplant breeding, but needed to be more strengthened for further developments in phylogenetic studies and mapping QTLs for agronomic traits.

Genetic transformation studies have been well established for production of several transgenic crops including eggplant. However, more efforts are required to utilize the full potential of genetic engineering in broad areas such as tolerance to multiple environmental stresses, improved nutritional quality and shelf life, delayed senescence, and postharvest quality for eggplant improvement. The emerging genome-editing techniques, particularly CRISPR/*Cas9* also needs to be more concentrated and further exploited to initiate efforts for precise and targeted breeding strategies, improving the nutritional quality of fruits, production of secondary metabolites, and introduction of agronomically useful traits for development of high yielding resistance varieties. *'Omics'* technology has opened new vistas for cost-efficient and time-saving sequencing techniques for wide applications in the areas of functional genomics, transcriptomics, proteomics and metabolomics. In the upcoming years, as the whole genome sequencing and transcriptome profiling of this non-model species become a routine job, the systematic efforts for improved fruit quality and yield-related traits in eggplants cultivars will definitely speed up to achieve the targeted breeding objectives.

REFERENCES

Agarwal M, Shrivastava N, Padh H, 2008. Advances in molecular marker techniques and their applications in plant sciences. *Plant Cell Reports* 27: 617-631.

Akanitapichat P, Phraibung K, Nuchklang K, Prompitakkul S, 2010. Antioxidant and hepatoprotective activities of five eggplant varieties. *Food and Chemical Toxicology* 48: 3017-3021.

Andersson HC, 2002. Calystegine alkaloids in Solanaceous food plants. Nordic Council of Ministers.

Andersson M, Turesson H, Olsson N, Fält AS, Ohlsson P, Gonzalez MN, Samuelsson M, Hofvander P, 2018. Genome editing in potato via CRISPRCas9 ribonucleoprotein delivery. *Physiologia Plantarum* 164: 378-384.

Ansari AM, Singh Y, Singh N, 2015. Marker based assessment of genetic diversity in aethiopicum and melongena species of genus Solanum. *Electronic Journal of Plant Breeding* 6: 904-917.

Arpaia S, Mennella G, Onofaro V, Perri E, Sunseri F, Rotino GL, 1997, Production of transgenic eggplant (*Solanum melongena* L.) resistant to Colorado Potato Beetle (*Leptinotarsa decemlineata* Say). Theoretical and Applied Genetics 95: 329-334.

Arumugam T, Anandakumar C, 2013. Genetic diversity in eggplant (*Solanum melongena* L.) *Plant Gene and Trait* 4.

Asano N, Kato A, Matsui K, Watson AA, Nash RJ, Molyneux RJ, Hackett L, Topping J, Winchester B, 1997. The effects of calystegines isolated from edible fruits and vegetables on mammalian liver glycosidases. *Glycobiology* 7: 1085-1088.

Barchi L, Lanteri S, Portis E, Acquadro A, Valè G, Toppino L, Rotino GL, 2011, Identification of SNP and SSR markers in eggplant using RAD tag sequencing. BMC Genomics 12: 1-9.

Barchi L, Lanteri S, Portis E, Valè G, Volante A, Pulcini L, Ciriaci T, Acciarri N, Barbierato V, Toppino L, Rotino GL, 2012, A RAD tag derived marker based eggplant linkage map and the location of QTLs determining anthocyanin pigmentation. *PloS One* 7: e43740.

Baysal Ö, Siragusa M, Gümrükcü E, Zengin S, Carimi F, Sajeva M, Da Silva JAT, 2010, Molecular characterization of *Fusarium oxysporum* f. *melongenae* by ISSR and RAPD markers on eggplant. *Biochemical Genetics* 48: 524-537.

Behera T, Sharma P, Singh B, Kumar G, Kumar R, Mohapatra T, Singh N, 2006, Assessment of genetic diversity and species relationships in eggplant (*Solanum melongena* L.) using STMS markers. *Scientia Horticulturae* 107: 352-357.

Bhatia S, Dahiya R, 2015, Plant-Based Biotechnological Products With Their Production Host, Modes of Delivery Systems, and Stability Testing. In: Modern

Applications of Plant Biotechnology in Pharmaceutical Sciences. Academic Press, Boston, pp 293-331.

Bhattacharyya P, Kumaria S, Kumar S, Tandon P, 2013, Start Codon Targeted (SCoT) marker reveals genetic diversity of *Dendrobium nobile* Lindl., an endangered medicinal orchid species. *Gene* 529: 21-26.

Carr PA, Church GM, 2009 Genome engineering. *Nature Biotechnology* 27: 1151.

Chen X, Zhang M, Tan J, Huang S, Wang C, Zhang H, Tan T, 2017, Comparative transcriptome analysis provides insights into molecular mechanisms for parthenocarpic fruit development in eggplant (*Solanum melongena* L.). *PLoS One* 12: e0179491.

Collard BC, Jahufer M, Brouwer J, Pang E, 2005, An introduction to markers, quantitative trait loci (QTL) mapping and marker-assisted selection for crop improvement: the basic concepts. *Euphytica* 142: 169-196.

Collard BC, Mackill DJ, 2009, Start codon targeted (SCoT) polymorphism: a simple, novel DNA marker technique for generating gene-targeted markers in plants. *Plant Molecular Biology Reporter* 27: 86.

Collonnier C, Mulya K, Fock I, Mariska I, Servaes A, Vedel F, Siljak-Yakovlev S, Souvannavong V, Ducreux G, Sihachakr D, 2001, Source of resistance against *Ralstonia solanacearum* in fertile somatic hybrids of eggplant (*Solanum melongena* L.) with *Solanum aethiopicum* L. *Plant Science* 160: 301-313.

Das M, Barua N, 2013, Pharmacological activities of *Solanum melongena* Linn. (Brinjal plant). *International Journal of Green Pharmacy (IJGP)* 7.

Daunay M, Lester R, 1988, The usefulness of taxonomy for Solanaceae breeders, with special reference to the genus Solanum and to *Solanum melongena* L. (eggplant).

Daunay MC, Chaput MH, Sihachakr D, Allot M, Vedel F, Ducreux G, 1993, Production and characterization of fertile somatic hybrids of eggplant (*Solanum melongena* L.) with *Solanum aethiopicum* L. *Theoretical and Applied Genetics* 85: 841-850.

Daunay MC, Laterrot H, Janick J, 2007, Iconography of the Solanaceae from antiquity to the Xviith Century: A rich source of information on genetic diversity and uses. In: International Society for Horticultural Science (ISHS), Leuven, Belgium, pp 59-88.

De Candolle A, 1885, Origin of cultivated plants. vol 48. D. Appleton,.

de Toledo Thomazella DP, Brail Q, Dahlbeck D, Staskawicz B, 2016, CRISPR-Cas9 mediated mutagenesis of a *DMR6* ortholog in tomato confers broad-spectrum disease resistance. *BioRxiv*: 064824.

Dempewolf H, Baute G, Anderson J, Kilian B, Smith C, Guarino L, 2017, Past and future use of wild relatives in crop breeding. *Crop Science* 57: 1070-1082.

Docimo T, Francese G, Ruggiero A, Batelli G, De Palma M, Bassolino L, Toppino L, Rotino GL, Mennella G, Tucci M, 2016, Phenylpropanoids accumulation in eggplant fruit: Characterization of biosynthetic genes and regulation by a MYB transcription factor. *Frontiers in Plant Science* 6.

Doбanlar S, Frary A, Daunay M-C, Huvenaars K, Mank R, Frary A, 2014, High resolution map of eggplant (*Solanum melongena*) reveals extensive chromosome rearrangement in domesticated members of the Solanaceae. *Euphytica* 198: 231-241.

Doganlar S, Frary A, Daunay M-C, Lester RN, Tanksley SD, 2002a, A comparative genetic linkage map of eggplant (*Solanum melongena*) and its implications for genome evolution in the Solanaceae. *Genetics* 161: 1697-1711.

Doganlar S, Frary A, Daunay M-C, Lester RN, Tanksley SD, 2002b, Conservation of gene function in the Solanaceae as revealed by comparative mapping of domestication traits in eggplant. *Genetics* 161: 1713-1726.

Donzella G, Spena A, Rotino GL, 2000, Transgenic parthenocarpic eggplants: superior germplasm for increased winter production. *Molecular Breeding* 6: 79-86.

El-Soda M, Malosetti M, Zwaan BJ, Koornneef M, Aarts MG, 2014, Genotype×environment interaction QTL mapping in plants: lessons from *Arabidopsis*. *Trends in Plant Science* 19: 390-398.

Enciso-Rodriguez F, Manrique-Carpintero NC, Nadakuduti SS, Buell CR, Zarka D, Douches D, 2019, Overcoming self-incompatibility in diploid potato using CRISPR-*Cas9*. *Frontiers in Plant Science* 10: 376.

Etminan A, Pour-Aboughadareh A, Mohammadi R, Ahmadi-Rad A, Noori A, Mahdavian Z, Moradi Z, 2016, Applicability of start codon targeted (SCoT) and inter-simple sequence repeat (ISSR) markers for genetic diversity analysis in durum wheat genotypes. *Biotechnology and Biotechnological Equipment* 30: 1075-1081.

Fassuliotis G, Nelson BV, Bhatt DP, 1981, Organogenesis in tissue culture of *Solanum melongena* cv. Florida Market. *Plant Science Letters* 22: 119-125.

Fischer R, Emans N, 2000, Molecular farming of pharmaceutical proteins. *Transgenic Research* 9: 279-299.

Fournier D, Lejeune F, Tourte Y, 1995, Cytological events during the initiation of meristematic nodules in calli derived from eggplant protoplasts. *Biology of the Cell* 85: 93-100.

Frary A, Doganlar S, Daunay M-C, Tanksley SD, 2003, QTL analysis of morphological traits in eggplant and implications for conservation of gene function during evolution of solanaceous species. *Theoretical and Applied Genetics* 107: 359-370.

Frary A, Frary A, Daunay M-C, Huvenaars K, Mank R, Doбanlar S, 2014, QTL hotspots in eggplant (*Solanum melongena*) detected with a high resolution map and CIM analysis. *Euphytica* 197: 211-228.

Fukuoka H, Miyatake K, Nunome T, Negoro S, Shirasawa K, Isobe S, Asamizu E, Yamaguchi H, Ohyama A, 2012, Development of gene-based markers and construction of an integrated linkage map in eggplant by using Solanum orthologous (SOL) gene sets. *Theoretical and Applied Genetics* 125: 47-56.

Furini A, Wunder J, 2004, Analysis of eggplant (*Solanum melongena*)-related germplasm: morphological and AFLP data contribute to phylogenetic interpretations and germplasm utilization. *Theoretical and Applied Genetics* 108: 197-208.

Gaj T, Gersbach CA, Barbas III CF, 2013, ZFN, TALEN, and CRISPR/Cas-based methods for genome engineering. *Trends in Biotechnology* 31: 397-405.

Galetto R, Duchateau P, Pâques F, 2009, Targeted approaches for gene therapy and the emergence of engineered meganucleases. *Expert Opinion on Biological Therapy* 9: 1289-1303.

Ge H, Liu Y, Jiang M, Zhang J, Han H, Chen H, 2013, Analysis of genetic diversity and structure of eggplant populations (*Solanum melongena* L.) in China using simple sequence repeat markers. *Scientia Horticulturae* 162: 71-75.

Gianoglio S, Moglia A, Comino C, Acquadro A, Jaime P, Diego O, Antonio G, Lanteri S, 2018, CRISPR/Cas9 knock-out of polyphenol oxidase genes for improving the quality of eggplant fruits. In: *Plant and Animal Genome XXVI*. PAG San Diego California USA, pp P1253-P1253.

Gleddie S, Keller W, Setterfield G, 1983, Somatic embryogenesis and plant regeneration from leaf explants and cell suspensions of *Solanum melongena* (eggplant). *Canadian Journal of Botany* 61: 656-666.

Gleddie S, Keller WA, Setterfield G, 1986, Production and characterization of somatic hybrids between *Solanum melongena* L. and *S. sisymbriifolium* Lam. *Theoretical and Applied Genetics* 71: 613-621.

Gleddiei S, Keller WA, Setterfield G, 1985, Plant regeneration from tissue, cell and protoplast cultures of several wild Solanum species. *Journal of Plant Physiology* 119: 405-418.

Gojko J, Sharon B, Qi C, James L, George H, Gerald G, 1998, Transformation of eggplant with synthetic *cryIIIA* gene produces a high level of resistance to the Colorado Potato Beetle. *Journal of the American Society for Horticultural Science (JASHS)* 123: 19-25.

Goldstein D, Thomas J, 2004, Biopharmaceuticals derived from genetically modified plants. *Qjm* 97: 705-716.

Gramazio P, Blanca J, Ziarsolo P, Herraiz F, Plazas M, Prohens J, Vilanova S, 2016, Transcriptome analysis and molecular marker discovery in *Solanum incanum* and *S. aethiopicum*, two close relatives of the common eggplant (*Solanum melongena*) with interest for breeding. *BMC Genomics* 17: 1-17.

Gramazio P, Prohens J, Borras D, Plazas M, Herraiz FJ, Vilanova S, 2017, Comparison of transcriptome-derived simple sequence repeat (SSR) and single nucleotide polymorphism (SNP) markers for genetic fingerprinting, diversity evaluation, and establishment of relationships in eggplants. *Euphytica* 213: 1-18.

Gramazio P, Prohens J, Plazas M, Andújar I, Herraiz FJ, Castillo E, Knapp S, Meyer RS, Vilanova S, 2014, Location of chlorogenic acid biosynthesis pathway and polyphenol oxidase genes in a new interspecific anchored linkage map of eggplant. BMC Plant Biology 14: 350.

Gramazio P, Yan H, Hasing T, Vilanova S, Prohens J, Bombarely A, 2019, Whole-genome resequencing of seven eggplant (*Solanum melongena*) and one wild relative (*S. incanum*) accessions provides new insights and breeding tools for eggplant enhancement. Frontiers in Plant Science 10: 1220.

Gürbüz N, Ului°ik S, Frary A, Frary A, Doбanlar S, 2018, Health benefits and bioactive compounds of eggplant. Food Chemistry 268: 602-610.

Guri A, Dunbar LJ, Sink KC, 1991, Somatic hybridization between selected *Lycopersicon* and Solanum species. Plant Cell Reports 10: 76-80.

Guri A, Sink KC, 1988, Organelle composition in somatic hybrids between an atrazine resistant biotype of *Solanum nigrum* and *Solanum melongena*. Plant Science 58: 51-58.

Guri A, Volokita M, Sink KC, 1987, Plant regeneration from leaf protoplasts of *Solanum torvum*. Plant Cell Reports 6: 302-304.

Hamilton GC, Jelenkovic GL, Lashomb JH, Ghidiu G, Billings S, Patt JM, 1997, Effectiveness of transgenic eggplant (*Solanum melongena* L.) against the Colorado Potato Beetle. Advances in Horticultural Science 11: 189-192.

Hirakawa H, Shirasawa K, Miyatake KO, Nunome T, Negoro S, Ohyama AK, Yamaguchi H, Sato S, Isobe S, Tabata S, Fukuoka H, 2014, Draft genome sequence of eggplant (*Solanum melongena* L.): the representative Solanum species indigenous to the old world. DNA Research 21: 649-660.

Hsu Patrick D, Lander Eric S, Zhang F, 2014, Development and applications of CRISPR-Cas9 for genome engineering. Cell 157: 1262-1278.

Hurtado M, Vilanova S, Plazas M, Gramazio P, Fonseka HH, Fonseka R, Prohens J, 2012, Diversity and relationships of eggplants from three geographically distant secondary centers of diversity. PLoS One 7: e41748.

Iannacone R, Grieco PD, Cellini F, 1997, Specific sequence modifications of a cry3B endotoxin gene result in high levels of expression and insect resistance. Plant Molecular Biology 34: 485-496.

Isshiki S, Iwata N, Khan MMR, 2008, ISSR variations in eggplant (*Solanum melongena* L.) and related Solanum species. Scientia Horticulturae 117: 186-190.

Isshiki S, Okubo H, Fujieda K, 1994, Phylogeny of eggplant and related Solanum species constructed by allozyme variation. Scientia Horticulturae 59: 171-176.

Isshiki S, Suzuki S, Yamashita K-i, 2003, RFLP analysis of mitochondrial DNA in eggplant and related Solanum species. Genetic Resources and Crop Evolution 50: 133-137.

Isshiki S, Uchiyama T, Tashiro Y, Miyazaki S, 1998, RFLP analysis of a PCR amplified region of chloroplast DNA in eggplant and related Solanum species. Euphytica 102: 295.

Jarl CI, Rietveld EM, de Haas JM, 1999, Transfer of fungal tolerance through interspecific somatic hybridisation between *Solanum melongena* and *S. torvum*. Plant Cell Reports 18: 791-796.

Kalloo G, Bergh B, 2012, Genetic improvement of vegetable crops. Newnes.

Kamat MG, Rao PS, 1978, Vegetative multiplication of eggplants (*Solanum melongena*) using tissue culture techniques. Plant Science Letters 13: 57-65.

Kardavan V, 2015, Molecular markers and their applications in plants. Journal of Plant Molecular Biology and Biotechnology 5: 7-12.

Karihaloo J, Brauner S, Gottlieb L, 1995, Random amplified polymorphic DNA variation in the eggplant, *Solanum melongena* L. (Solanaceae). Theoretical and Applied Genetics 90: 767-770.

Karihaloo J, Gottlieb L, 1995, Allozyme variation in the eggplant, *Solanum melongena* L. (Solanaceae). Theoretical and Applied Genetics 90: 578-583.

Karihaloo J, Kaur M, Singh S, 2002, Seed protein diversity in *Solanum melongena* L. and its wild and weedy relatives. Genetic Resources and Crop Evolution 49: 533-539.

Kashyap V, Kumar SV, Collonnier C, Fusari F, Haicour R, Rotino GL, Sihachakr D, Rajam MV, 2003, Biotechnology of eggplant. Scientia Horticulturae 97: 1-25.

Kaushik P, 2019, Genetic analysis for fruit phenolics content, flesh color, and browning related traits in eggplant (*Solanum melongena*). International Journal of Molecular Sciences 20: 2990.

Kearsey M, Farquhar A, 1998, QTL analysis in plants; where are we now?. Heredity 80: 137-142.

Kirti P, Moorty K, Rao S, Rao BS, 1984, Cytological observations on some autotetraploids and amphidiploids in spinous solanums and their bearing on interrelation'ships. Current Science 53: 1256-1258.

Knapp S, Vorontsova MS, Prohens J, 2013, Wild relatives of the eggplant (*Solanum melongena* L.: Solanaceae): new understanding of species names in a complex group. PloS One 8: e57039.

Koley TK, Tiwari SK, Sarkar A, Nishad J, Goswami A, Singh B, 2019, Antioxidant potential of indian eggplant: Comparison among white, purple and green genotypes using chemometrics. Agricultural Research 8: 9-20.

Koundal M, Sharma D, Mohapatra T, Koundal K, 2006, Comparative evaluation of RAPD.and AFLP based genetic diversity in Brinjal (*Solanum melongena*). Journal of Plant Biochemistry and Biotechnology 15: 15-19.

Kowalczyk TP, Mackenzie IA, Cocking EC, 1983, Plant regeneration from organ expiants and protoplasts of the medicinal plant *Solatium khasianum* C. B. Clarke var. chatterjeeanum Sengupta (Syn. *Solatium viarum* Dunal). Zeitschrift für Pflanzenphysiologie 111: 55-68.

Kumar PA, Mandaokar A, Sreenivasu K, Chakrabarti SK, Bisaria S, Sharma SR, Kaur S, Sharma RP, 1998, Insect-resistant transgenic brinjal plants. Molecular Breeding 4: 33-37.

Lebeau A, Gouy M, Daunay MC, Wicker E, Chiroleu F, Prior P, Frary A, Dintinger J, 2013, Genetic mapping of a major dominant gene for resistance to *Ralstonia solanacearum* in eggplant. Theoretical and Applied Genetics 126: 143-158.

Lester R, Hasan S, 1991, Origin and domestication of the brinjal eggplant, *Solanum melongena,* from *S. incanum,* in Africa and Asia Hawkes, J, G, Lester, R, N, Nee, M, Estrada, N ed (s) Solanaceae III Taxonomy, chemistry, evolution Roy Bot Gard: Kew and Linnean Soc: London: 369-387.

Li J, Gao Z, Zhou L, Li L, Zhang J, Liu Y, Chen H, 2019, Comparative transcriptome analysis reveals K+ transporter gene contributing to salt tolerance in eggplant. BMC Plant Biology 19: 1-18.

Li J, He Y-J, Zhou L, Liu Y, Jiang M, Ren L, Chen H, 2018, Transcriptome profiling of genes related to light-induced anthocyanin biosynthesis in eggplant (*Solanum melongena* L.) before purple color becomes evident. BMC Genomics 19: 1-12.

Li J, Ren L, Gao Z, Jiang M, Liu Y, Zhou L, He Y, Chen H, 2017, Combined transcriptomic and proteomic analysis constructs a new model for lightinduced anthocyanin biosynthesis in eggplant (*Solanum melongena* L.). Plant, Cell and Environment 40: 3069-3087.

Liu KB, Li YM, Sink KC, 1995, Asymmetric somatic hybrid plants between an interspecific *Lycopersicon* hybrid and *Solanum melongena*. Plant Cell Reports 14: 652-656.

Liu Y, Yin X, Sun YP, Liu Y, Zhou YY, Pan J, Guan W, Yang BY, Kuang HX, 2020, Chemical constituent from the roots of *Solanum melongena* L. and their potential anti-inflammatory activity. Natural Product Research: 1-8.

Luo C, He X-H, Chen H, Hu Y, Ou S-J, 2012, Genetic relationship and diversity of *Mangifera indica* L.: revealed through SCoT analysis. Genetic Resources and Crop Evolution 59: 1505-1515.

Ma JK, Drake PM, Christou P, 2003, The production of recombinant pharmaceutical proteins in plants. Nature Reviews Genetics 4: 794-805.

Magioli C, Mansur E, 2005, Eggplant (*Solanum melongena* L.): tissue culture, genetic transformation and use as an alternative model plant. Acta Botanica Brasilica 19: 139-148.

Mahjbi A, Baraket G, Oueslati A, Salhi-Hannachi A, 2015, Start Codon Targeted (SCoT) markers provide new insights into the genetic diversity analysis and characterization of Tunisian Citrus species. Biochemical Systematics and Ecology 61: 390-398.

Maioli A, Gianoglio S, Moglia A, Acquadro A, Valentino D, Milani AM, Prohens J, Orzaez D, Granell A, Lanteri S, Comino C, 2020, Simultaneous CRISPR/Cas9 editing of three PPO genes reduces fruit flesh browning in *Solanum melongena* L. Frontiers in Plant Science 11: 1883.

Matsuoka H, Hinata K, 1979, NAA-induced organogenesis and embryogenesis in hypocotyl callus of *Solarium melongena* L. Journal of Experimental Botany 30: 363-370.

McCammon K, Honma S, 1983, Morphological and cytogenetic analyses of an interspecific hybrid eggplant, *Solanum melongena* X *Solanum torvum*.

Meyer RS, Bamshad M, Fuller DQ, Litt A, 2014, Comparing medicinal uses of eggplant and related Solanaceae in China, India, and the Philippines suggests the independent development of uses, cultural diffusion, and recent species substitutions. Economic Botany 68: 137-152.

Mishiba KI, Nishida K, Inoue N, Fujiwara T, Teranishi S, Iwata Y, Takeda S, Koizumi N, 2020, Genetic engineering of eggplant accumulating B-carotene in fruit. Plant Cell Reports 39: 1029-1039.

Mishra P, Tiwari SK, Kashyap SP, Tiwari KN, Singh M, Singh B, 2020, High-density genetic linkage map based on arbitrary and microsatellite markers using inter-specific recombinant inbred lines in eggplant (*Solanum melongena* L.) Journal of Plant Biochemistry and Biotechnology 29: 427-438.

Miyatake K, Saito T, Negoro S, Yamaguchi H, Nunome T, Ohyama A, Fukuoka H, 2012, Development of selective markers linked to a major QTL for parthenocarpy in eggplant (*Solanum melongena* L.). Theoretical and Applied Genetics 124: 1403-1413.

Miyatake K, Saito T, Negoro S, Yamaguchi H, Nunome T, Ohyama A, Fukuoka H, 2016, Detailed mapping of a resistance locus against Fusarium wilt in cultivated eggplant (*Solanum melongena*). Theoretical and applied genetics 129: 357-367.

Miyatake K, Saito T, Nunome T, Yamaguchi H, Negoro S, Ohyama A, Wu J, Katayose Y, Fukuoka H, 2020, Fine mapping of a major locus representing the lack of prickles in eggplant revealed the availability of a 0.5-kb insertion/deletion for marker-assisted selection. Breeding Science: 20004.

Mulpuri S, Muddanuru T, Francis G, 2013, Start codon targeted (SCoT) polymorphism in toxic and non-toxic accessions of *Jatropha curcas* L. and development of a codominant SCAR marker. Plant Science 207: 117-127.

Mutalik S, Paridhavi K, Rao CM, Udupa N, 2003, Antipyretic and analgesic effect of leaves of *Solanum melongena* Linn. in rodents. Indian Journal of Pharmacology 35: 312-315.

Mutlu N, Boyaci FH, Göçmen M, Abak K, 2008, Development of SRAP, SRAP-RGA, RAPD and SCAR markers linked with a Fusarium wilt resistance gene in eggplant. Theoretical and applied genetics 117: 1303-1312.

Nunome T, Ishiguro K, Yoshida T, Hirai M, 2001, Mapping of fruit shape and color development traits in eggplant (*Solanum melongena* L.) based on RAPD and AFLP markers. Breeding Science 51: 19-26.

Nunome T, Negoro S, Kono I, Kanamori H, Miyatake K, Yamaguchi H, Ohyama A, Fukuoka H, 2009, Development of SSR markers derived from SSR-enriched genomic library of eggplant (*Solanum melongena* L.). Theoretical and Applied Genetics 119: 1143-1153.

Nunome T, Suwabe K, Iketani H, Hirai M, Wricke G, 2003, Identification and characterization of microsatellites in eggplant. Plant Breeding 122: 256-262.

Nunome T, Yoshida T, Hirai M, 1998, Genetic linkage map of eggplant. In: Proceedings of the 10th Eucarpia meeting on genetics and breeding of capsicum and eggplant, Avignon France, pp 239-242.

Obembe OO, Popoola JO, Leelavathi S, Reddy SV, 2011, Advances in plant molecular farming. Biotechnology Advances 29: 210-222.

Pâques F, Duchateau P, 2007, Meganucleases and DNA double-strand break-induced recombination: perspectives for gene therapy. Current Gene Therapy 7: 49-66.

Paran I, Zamir D, 2003, Quantitative traits in plants: beyond the QTL. Trends in Genetics 19: 303-306.

Paterson AH, 2002, What has QTL mapping taught us about plant domestication?. New Phytologist 154: 591-608.

Plazas M, Andujar I, Vilanova S, Hurtado M, Gramazio P, Herraiz FJ, Prohens J, 2013, Breeding for chlorogenic acid content in eggplant: interest and prospects. Notulae Botanicae Horti Agrobotanici Cluj-Napoca 41: 26-35.

Portis E, Barchi L, Toppino L, Lanteri S, Acciarri N, Felicioni N, Fusari F, Barbierato V, Cericola F, Valè G, Rotino GL, 2014, QTL mapping in eggplant reveals clusters of yield-related loci and orthology with the tomato genome. PLoS One 9: e89499.

Portis E, Cericola F, Barchi L, Toppino L, Acciarri N, Pulcini L, Sala T, Lanteri S, Rotino GL, 2015, Association mapping for fruit, plant and leaf morphology traits in eggplant. PLoS One 10: e0135200.

Portis E, Lanteri S, Barchi L, Portis F, Valente L, Toppino L, Rotino GL, Acquadro A, 2018, Comprehensive characterization of simple sequence repeats in eggplant (*Solanum melongena* L.) genome and construction of a web resource. Frontiers in Plant Science 9: 401.

Prabhavathi V, Yadav JS, Kumar PA, Rajam MV, 2002, Abiotic stress tolerance in transgenic eggplant (*Solanum melongena* L.) by introduction of bacterial mannitol phosphodehydrogenase gene. Molecular Breeding 9: 137-147.

Rajput ZI, Hu S-h, Xiao C-w, Arijo AG, 2007, Adjuvant effects of saponins on animal immune responses. J Zhejiang Univ Sci B 8: 153-161.

Ramesh KR, Hemalatha R, Vijayendra CA, Arshi UZS, Dushyant SB, Dinesh KB, 2016, Transcriptome analysis of *Solanum melongena* L. (eggplant) fruit to identify putative allergens and their epitopes. Gene 576: 64-71.

Ran FA, Hsu PD, Wright J, Agarwala V, Scott DA, Zhang F, 2013, Genome engineering using the CRISPR-Cas9 system. Nature Protocols 8: 2281-2308.

Rao G, Baksh S, 1979, Cytomorphological study of the amphidiploids derived from the hybrids of the crosses between *Solanum melongena* L. and *Solanum integrifolium* Poir. Current Science.

Rao PS, Narayanaswami S, 1968, Induced morphogenesis in tissue cultures of *Solanum xanthocarpum*. Planta 81: 372-375.

Redondo P, Prieto J, Munoz IG, Alibés A, Stricher F, Serrano L, Cabaniols JP, Daboussi F, Arnould S, Perez C, Duchateau P, 2008, Molecular basis of xeroderma pigmentosum group C DNA recognition by engineered meganucleases. Nature 456: 107-111.

Rinaldi R, Van Deynze A, Portis E, Rotino GL, Toppino L, Hill T, Ashrafi H, Barchi L, Lanteri S, 2016, New insights on eggplant/tomato/pepper synteny and identification of eggplant and pepper orthologous QTL. Frontiers in plant science 7: 1031.

Rizza F, Mennella G, Collonnier C, Sihachakr D, Kashyap V, Rajam M, Prestera M, Rotino G, 2002, Androgenic dihaploids from somatic hybrids between *Solanum melongena* and *S. aethiopicum* group gilo as a source of resistance to *Fusarium oxysporum* f. sp. *melongenae*. Plant Cell Reports 20: 1022-1032.

Rotino GL, Perri E, Zottini M, Sommer H, Spena A, 1997, Genetic engineering of parthenocarpic plants. Nature Biotechnology 15: 1398-1401.

Saini DK, Kaushik P, 2019, Visiting eggplant from a biotechnological perspective: A review. Scientia Horticulturae 253: 327-340.

Sakata Y, Lester RN, 1994, Chloroplast DNA diversity in eggplant (*Solanum melongena*) and its related species *S. incanum* and *S. marginatum*. Euphytica 80: 1-4.

Sakata Y, Nishio T, Matthews PJ, 1991, Chloroplast DNA analysis of eggplant (*Solanum melongena*) and related species for their taxonomic affinity. Euphytica 55: 21-26.

Salgon S, Jourda C, Sauvage C, Daunay M-C, Reynaud B, Wicker E, Dintinger J, 2017, Eggplant resistance to the *Ralstonia solanacearum* species complex involves both broad-spectrum and strain-specific quantitative trait loci. Frontiers in Plant Science 8.

Satya P, Karan M, Jana S, Mitra S, Sharma A, Karmakar PG, Ray DP, 2015, Start codon targeted (SCoT) polymorphism reveals genetic diversity in wild and domesticated populations of ramie (*Boehmeria nivea* L. Gaudich.), a premium textile fiber producing species. Meta Gene 3: 62-70.

Shahlaei A, Torabi S, Khosroshahli M, 2014, Efficiacy of SCoT and ISSR marekers in assesment of tomato (*Lycopersicum esculentum* Mill.) genetic diversity. International Journal of Biosciences 5: 14-22.

Sharma D, 1980, Interspecific hybridization in genus Solanum. A cross between *S. melongena* and *S. khasianum* through embryo culture.

Sharma D, Sareen P, Chowdhury J, 1984, Crossability and pollination in some non-tuberous Solanum species. Indian Journal of Agricultural Sciences.

Sharon B, Gojko J, Chee-Kok C, Jodi E, 1997, The effect of growth regulators and antibiotics on eggplant transformation. Journal of the American Society for Horticultural Science 122: 158-162.

Sihachakr D, Ducreux G, 1987, Cultural behavior of protoplasts from different organs of eggplant (*solanum melongena* L.), and plant regeneration. Plant Cell, Tissue and Organ Culture 11: 179-188.

Silva G, Poirot L, Galetto R, Smith J, Montoya G, Duchateau P, Pâques F, 2011, Meganucleases and other tools for targeted genome engineering: perspectives and challenges for gene therapy. Current Gene Therapy 11: 11-27.

Singh A, Singh R, Kumar S, Kalloo G, 2006, Genetic diversity within the genus Solanum (Solanaceae) as revealed by RAPD markers. Current Science: 711-716.

Singh A, Verma S, Bansal K, 2010, Plastid transformation in eggplant (*Solanum melongena* L.). Transgenic Research 19: 113-119.

Singh Yadav J, Venkat Rajam M, 1998, Temporal Regulation of somatic embryogenesis by adjusting cellular polyamine content in eggplant. Plant Physiology 116: 617.

Sreekar K, 2020, Biotechnology and its implications in brinjal improvement: A review. Journal of Pharmacognosy and Phytochemistry 9: 1096-1102.

Stuber CW, Polacco M, Senior ML, 1999, Synergy of empirical breeding, marker-assisted selection, and genomics to increase crop yield potential. Crop Science 39: 1571-1583.

Sciancalepore A, Martelli G, Acciarri N, Rotino GL, Valentino D, Tamietti G, Sunseri F, 2003, Development of RAPD-AFLP map of eggplant and improvement of tolerance to verticillium wilt. In: XXVI International Horticultural Congress: Biotechnology in Horticultural Crop Improvement: Achievements, Opportunities pp. 107-115.

Thapa SP, Miyao EM, Michael Davis R, Coaker G, 2015, Identification of QTLs controlling resistance to *Pseudomonas syringae* pv. tomato race 1 strains from the wild tomato, *Solanum habrochaites* LA1777. Theoretical and Applied Genetics 128: 681-692.

Tiwari SK, 2007, Development and utilization of microsatellite markers in germplasm characterization of brinjal (*Solanum melongena* L.): Thesis (Ph. D). IARI, Division of Plant Genetic Resources.

Tiwari SK, Karihaloo J, Hameed N, Gaikwad AB, 2009, Molecular characterization of brinjal (*Solanum melongena* L.) cultivars using RAPD and ISSR markers. Journal of Plant Biochemistry and Biotechnology 18: 189-195.

Toki S, Kameya T, Abe T, 1990, Production of a triple mutant, chlorophyll-deficient, streptomycin-, and kanamycin-resistant *Nicotiana tabacum*, and its use in intergeneric somatic hybrid formation with *Solanum melongena*. Theoretical and Applied Genetics 80: 588-592.

Toppino L, Barchi L, Lo Scalzo R, Palazzolo E, Francese G, Fibiani M, D'Alessandro A, Papa V, Laudicina VA, Sabatino L, Pulcini L, 2016, Mapping quantitative trait loci affecting biochemical and morphological fruit properties in eggplant (*Solanum melongena* L.). Frontiers in Plant Science 7: 256.

Toppino L, Barchi L, Mercati F, Acciarri N, Perrone D, Martina M, Gattolin S, Sala T, Fadda S, Mauceri A, Ciriaci T, 2020, A new intra-specific and high-resolution genetic map of eggplant based on a RIL population, and location of QTLs related to plant anthocyanin pigmentation and seed vigour. Genes 11: 745.

Ueta R, Abe C, Watanabe T, Sugano SS, Ishihara R, Ezura H, Osakabe Y, Osakabe K, 2017, Rapid breeding of parthenocarpic tomato plants using CRISPR/Cas9. Scientific Reports 7: 1-8.

van der Weerden GM, Barendse GW, 2006, A web-based searchable database developed for the EGGNET project and applied to the Radboud University Solanaceae database. In: VI International Solanaceae Conference: Genomics Meets Biodiversity 745, 503-506.

Van Eck J, 2018, Genome editing and plant transformation of Solanaceous food crops. Current Opinion in Biotechnology 49: 35-41.

Varshney RK, Chabane K, Hendre PS, Aggarwal RK, Graner A, 2007, Comparative assessment of EST-SSR, EST-SNP and AFLP markers for evaluation of genetic diversity and conservation of genetic resources using wild, cultivated and elite barleys. Plant Science 173: 638-649.

Vavilov NI, 1951, The origin, variation, immunity and breeding of cultivated plants. LWW 72, 6.

Venora G, Russo C, Errico, 1992, A karyotype analysis in *Solanum melongena* L. In: Eucarpia Proceedings of the 7th Meeting on Genetics and breeding of Capsicum and Eggplant, 7-10.

Wang L, Li J, Zhao J, He C, 2015a, Evolutionary developmental genetics of fruit morphological variation within the Solanaceae. Frontiers in Plant Science 6: 248.

Wang S, Zhang S, Wang W, Xiong X, Meng F, Cui X, 2015b, Efficient targeted mutagenesis in potato by the CRISPR/Cas9 system. Plant Cell Reports 34: 1473-1476.

Wang Y, Liu A, Li W, Jiang Y, Song S, Li Y, Chen R, 2017, Comparative proteomic analysis of eggplant (*Solanum melongena* L.) heterostylous pistil development. PLoS One 12: e0179018.

Wei Q, Du L, Wang W, Hu T, Hu H, Wang J, David K, Bao C, 2019, Comparative transcriptome analysis in eggplant reveals selection trends during eggplant domestication. International Journal of Genomics 9: 2019.

Wei Q, Wang J, Wang W, Hu T, Hu H, Bao C, 2020a, A high-quality chromosome-level genome assembly reveals genetics for important traits in eggplant. Horticulture Research 7: 1-15.

Wei Q, Wang W, Hu T, Hu H, Wang J, Bao C, 2020b, Construction of a SNP-based genetic map using SLAF-Seq and QTL analysis of morphological traits in eggplant. Frontiers in Genetics 11: 178.

Wu F, Eannetta NT, Xu Y, Tanksley SD, 2009a, A detailed synteny map of the eggplant genome based on conserved ortholog set II (COSII) markers. Theoretical and Applied Genetics 118: 927-935.

Wu J, Gutierrez OA, Jenkins JN, McCarty JC, Zhu J, 2009b, Quantitative analysis and QTL mapping for agronomic and fiber traits in an RI population of upland cotton. Euphytica 165: 231-245.

Xiong F, Zhong R, Han Z, Jiang J, He L, Zhuang W, Tang R, 2011, Start codon targeted polymorphism for evaluation of functional genetic variation and relationships in cultivated peanut (*Arachis hypogaea* L.) genotypes. Molecular Biology Reports 38: 3487-3494.

Yang B-Y, Yin X, Liu Y, Sun Y, Guan W, Zhou Y-Y, Kuang H-X, 2020, Terpenes and lignans from the roots of *Solanum melongena* L. Natural Product Research 34: 359-368.

Yang X, Cheng Y-F, Deng C, Ma Y, Wang Z-W, Chen X-H, Xue L-B, 2014, Comparative transcriptome analysis of eggplant (*Solanum melongena* L.) and turkey berry (*Solanum torvum* Sw.): phylogenomics and disease resistance analysis. BMC Genomics 15: 1-13.

Yang X, Zhang Y, Cheng Y, Chen X, 2019, Transcriptome analysis reveals multiple signal network contributing to the verticillium wilt resistance in eggplant. Scientia Horticulturae 256: 108576.

Yang Y, Bao S, Zhou X, Liu J, Zhuang Y, 2018, The key genes and pathways related to male sterility of eggplant revealed by comparative transcriptome analysis. BMC Plant Biology 18: 1-13.

Zhang A, Zhu Z, Shang J, Zhang S, Shen H, Wu X, Zha D, 2020, Transcriptome profiling and gene expression analyses of eggplant (*Solanum melongena* L.) under heat stress. PLoS One 15: e0236980.

Zhang H, Mittal N, Leamy LJ, Barazani O, Song B-H, 2017, Back into the wild-Apply untapped genetic diversity of wild relatives for crop improvement. Evolutionary Applications 10: 5-24.

Zhang S, Zhang A, Wu X, Zhu Z, Yang Z, Zhu Y, Zha D, 2019, Transcriptome analysis revealed expression of genes related to anthocyanin biosynthesis in eggplant (*Solanum melongena* L.) under high-temperature stress. BMC Plant Biology 19: 1-13.

Zhi J, Liu X, Li D, Huang Y, Yan S, Cao B, Qiu Z, 2020, CRISPR/Cas9-mediated *SlAN2* mutants reveal various regulatory models of anthocyanin biosynthesis in tomato plant. Plant Cell Reports 39: 799-809

Chapter 12

TILLING–The Art of Gene Hunt: Recent Advances in Vegetable Improvement through Mutation and Gene Discovery

J. Jyothsna[1], Reena Nair[2] and A.K. Mehta[3]

[1]Ph.D. Scholar, Department of Horticulture,
[2]Assistant Professor, Department of Horticulture,
[3]University Professor and Head, Department of Plant Breeding and Genetics,
College of Agriculture, Jawaharlal Nehru Krishi Vishwavidyalaya, Madhya Pradesh

Mutations, either spontaneous or induced, can contribute to new alleles that are in assortment with desirable phenotypic traits; are used in the process of basic researches investigating the gene functions. The identification of mutations earlier was limited as they are circumscribed to the phenotypic expression. The recent developments in genome sequencing of many important crops and emerging molecular techniques have greatly eased the discovery of mutations in the desirable genes. The TILLING strategy is one such an outcome of increased knowledge on biotechnological applications and development in molecular techniques. The TILLING technique is a powerful tool of reverse genetics approach that pairs the chemically induced mutagenesis with high-throughput screening of individuals carrying a mutation in the genes of interest. The application of the TILLING technique in breeding programs and functional genomics of vegetable crops is apprised in this chapter.

Introduction

The concept of functional genomics embraces several genetic approaches directing to the determination of gene function on a genome-wide scale. These approaches are favored by the application of new, high-throughput technologies that help to tap the genetic information and functions from any organism. The techniques of functional genomics cleave into forward and reverse genetic approaches, based on the germ that leads to advanced investigation, either the phenotype or the gene.

The forward genetics is an orthodox approach that studies the gene function through examining the phenotype (of the mutant plant), proceeding to the gene underlying the phenotype. Whilst, the strategies of forward genetics are greatly useful to infer important features of a novel gene from the slant of biochemical and molecular aspects, they are realized as less-informative as for the functional complexity of a living organism and barely meet the requirement of high-throughput and large scale survey of gene functions, besides most phenotypes are unclear.

Different reverse genetic approaches, wherein the investigation begins from the genotype (a DNA sequence) and go forward to the phenotype by creating alterations to the genotype were conceptualized to subdue the limitations of forward genetic approaches. The technological advances in the reverse genetics and computational tools have rendered massive genomic resources, which have ensued identification and annotation of thousands of novel genes for crop improvement. The Targeting Induced Local Lesions in Genomes, (TILLING) is one such a novel reverse genetics strategy that utilizes the vantages of classical mutagenesis and the high throughput method of mutation detection. The below sections of this chapter aims at a clear understanding of how the TILLING technique facilitates crop improvement through mutation breeding by enabling the discovery of mutations in the gene of interest.

Role of TILLING in Mutation Breeding Programs

The mutagenesis of plant cells with physical and chemical mutagens is a well-established strategy for crop improvement. The mutations are more often employed to improve a single trait in an elite background. The easy-to-manifest traits like disease resistance, earliness are fundamentally targeted in the mutation improvement programs. On the other hand, mutants expressing a deleterious phenotype are normally discarded by the plant breeder perhaps it may provide an opportunity for the molecular understanding of the phenotypic lesions. A typical example is the fate of a *xantha* mutant, a chlorophyll deficient mutant that is discarded in the seedling stage (Von *et al.*, 1974). The cloning and investigating the genes associated with such mutations are likely to offer new insights into crop improvement and provide a better understanding of molecular aspects related to such mutations.

From above, it can be corroborated how mutagenesis, and gene discovery and their identification through TILLING can step up the crop improvement programs. Researches have been focused on combining mutagenesis, TILLING, and functional genomics research that can intercept unidentified novel gene mutants with crop improvement. Recent developments in TILLING, since 2000, has subscribed to the revival of chemical mutagenesis in reverse genetics due to the efficiency of the new techniques.

Basic Concepts of the TILLING Technique

TILLING is a novel approach of reverse genetics that has emerged into a successful tool for gene discovery, reveling through the advantages of point mutation generated by the chemical mutagens and PCR based mutational screening. It is increasingly recognized as a high-throughput technique that enables rapid and budget detection of the induced point mutation in the chemically mutagenized population. Its applicability to any organism besides the model organisms, high sensitivity, high efficiency are the interesting features that have empowered TILLING into a potential tool for gene discovery, DNA polymorphism assessment at the single nucleotide level and ultimately crop improvement.

Some plant species are likely to produce highly desirable traits by knocking out a gene, but lacking the practical reverse genetics application. In such cases, TILLING can locate an allelic series of induced point mutations – missense mutation and truncation lesions, with advanced gene knockout methods. The interest is this which made the TILLING an attractive technique to fit a huge range application from basic functional genomic study to practical crop breeding.

The technique of TILLING was first described by Claire McCallum and colleagues at the Fred Hutchinson Cancer Research Center in Seattle, WA, in the 1990s. The doability of TILLING was realized in a study demonstrating the detection of two chromomethylase (CMT) genes of *Arabidopsis* mutagenized with the traditional chemical mutagen (McCallum *et al.*, 2000). The technique was developed for *Arabidopsis* as a substitute for insertional mutagenesis. Subsequently, its application was broadcasted to a wide range of plant species, including agronomically important crops such as barley, wheat, and maize (Kurowska *et al.*, 2012). The very objective of TILLING is to identify individuals bearing mutations in the gene of interest and to assign a function to the analyzed gene through phenotypic analyses of the identified mutants. Not only to the unidentified genes, but TILLING can also be executed on genes of known identity and function to produce a series of alleles expressing modification in the trait of interest, with the concern that the newly identified alleles can successfully be used as a valuable genome resource in breeding programs.

Discovery of Mutations in the Gene of Interest through TILLING: The Procedure

The TILLING protocol broadly involves four main phases.

1. Generation of a mutant population
2. Selection of target genes
3. Molecular screening
4. Recovery of mutants

Generation of Mutant Populations

The Prerequisite

Well-Planned programming of the chemical mutagenesis drives the subsequent experimental choice and quality of TILLING result into a success. The indispensable element of TILLING action is to generate an appropriate frequency of mutations per genome along with maintaining a pretty low lethality and plant sterility percentage. Before the commencement of the experiment, two basic aspects need to be deliberated.

i. Choice of mutagen
ii. Material to be mutagenized

For TILLING purpose, generally, the chemical mutagens - ethyl methane sulphonate (EMS), N-methyl-N-nitrosourea (MNU), or sodium azide (NaN_3, SA), are utilized. For technical ease, the seed is widely mutagenized, although pollen and other vegetative parts can also be subjected to mutagenic treatment. The selection of a cultivar or a line is objective-dependent in line with the purpose of the investigation.

Why Often Chemical Mutagens?

While the physical mutagens like gamma rays, thermal neutrons and X-rays have bestowed their best to the development of numerous mutant varieties in plant species, the chemical mutagens are highly favored, as far as TILLING is concerned. This bias is why because the chemical mutagens, mostly the alkylating agents, directly alter the structure and properties of DNA bases, having the potential of generating every possible type of mutation *viz.*, transitions, transversions, deletions, and frameshifts along with a relatively low frequency of the chromosome breakages that cause various chromosomal rearrangements (Maluszynski *et al.*, 2003). Of all, they induce irreversible point mutation in relatively high densities. Moreover, they generate an allelic series at any target locus resulting in a decline in the abnormalities in protein function and these changes are worthwhile for assigning a gene, a function.

Chemical Mutagens and the Somatic Effect

As mentioned earlier, the alkylating agents - EMS, NMS, and SA are widely used for creating a TILLING population. The potential of the chemical mutagen to generate the optimum frequency of point mutation per genome is preponderant. The desirability of a mutagen depends on the number of mutations it induces per genome. Among the chemical mutagens, EMS is the commonest mutagen entertained over several decades.

EMS offers advantages such as the relatively higher density of irreversible mutations per genome. NMU is reported as a very strong chemical mutagen and at times called 'supermutagen' (Maluszynski *et al.*, 2009). SA has proved to be a potential mutagen in microorganisms[11] yet an underexploited mutant in most crop species. The alkylating agents such as EMS and MNU induce GC>AT transitions (Maluszynski *et al.*, 2003; Till *et al.*, 2003). They cause changes in DNA by transferring a methyl (–CH_3) group to the oxygen and nitrogen atoms of the nucleotide bases (Warren *et al.*, 2003).

The right dose of the chemical mutagen is also a crucial factor to obtain the right level of mutation. Even though the higher doses increase the frequency of point mutation, they are least desirable perhaps undesirable mutations can also be resulted. Too high dosage leads to high lethality and too low dosage in low mutation density, subsequently increasing the costs and labor required for managing a larger mutagenized population. Therefore, it is recommended to entertain a dosage that induces the highest frequency of point mutation in M_2 plants, irrespective of the somatic effect in M_1 plants (Martin *et al.*, 2009).

In addition to the DNA lesions in the nucleus and cytoplasmic organelles, mutagens also creates damages and disturbances to the cytosol and cell cycle, respectively, leading to the deterioration of cell metabolisms and are called as the 'somatic effect of the mutagen' which are manifested as delay in seed germination, reduction in plant emergence, growth retardation, appearance of chlorophyll defects, reduced fertility and plant survival (Szarejko *et al.*, 2011).

Mutagenic Treatment

There are four steps in the mutagenic treatment of seed with chemical mutagens.

i. Presoaking the seeds in distilled water
ii. Treatment with the chemical mutagen
iii. Post-treatment rinsing with water
iv. Planting M_1 seeds

Besides the dosage of mutagen, the duration of the treatment is also a crucial factor, deserving equivalent importance, and role in influencing the mutation. The seeds treated with an elevated dosage for a prolonged period are likely to suffer more from the somatic effect of the mutagen which is caused by the inability of the chemical mutagen to diffuse uniformly into all cells of the seed material (Bovina *et al.*, 2011).

Before proceeding to the treatment with chemical mutagens, the seeds are generally soaked in distilled water to physiologically activate the seed and cut down the somatic effect of the mutagen. Essentially an extensive post-treatment rinsing of the treated seed with an ample amount of water is to be done to remove its residues from seed surface.

Handling of the Mutated Populations

The seeds subjected to the mutagenic treatment can preferably be raised in a greenhouse or a growth chamber. For easy identification and handling, each M_1 plant to be assigned with a number. A large population of about 10,000 plants in M_1 generation serves the purpose of TILLING. At maturity, seeds from M_1 plants are harvested and M_2 seed bank is established. To avoid the repetitions of the same mutations, only one M_2 individual must be forwarded from each M_1 plant. The M_3 plants are derived from seeds of the M_2 plants. A gist on the mutant population/platform generated in vegetable crops is given in Table 12.1.

Selection of Target Genes

Once the TILLING population is developed, the subsequent action to be taken is the selection of the gene of interest, the target gene. Typically, the selected gene(s) of a known sequence has a putative involvement in a particular biological or physiological process (Bovina *et al.*, 2011). The single-copy genes are considered for TILLING as the point mutations incline to generate easily detectable phenotypic alterations.

Table 12.1: Mutant Population/Platform Created in Vegetable for TILLING

Sl.No.	Crop	Parent/Donor line	Mutagen	Reference
1	Tomato	cv. Micro-Tom	EMS	Baldet *et al.* (2013)
2		AJ 222798, AF029984	EMS	Jones *et al.* (2012)
3		cv. Micro-Tom	EMS	Just *et al.* (2013)
4		cv. M2,cv. Red Setter	EMS	Minoia *et al.* (2016)
5		cv. M82	EMS	Piron *et al.* (2010)
6		cv. Micro-Tom	EMS	Pulungan *et al.* (2018)
7		cv. Red Setter	EMS	Silletti *et al.* (2013)
8	Cucumber	cv. Biet Alpha	EMS	Boualem *et al.* (2019)
9		Inbred line	EMS	Chen *et al.* (2018)
10	Musk Melon	cv. CharMono	EMS	Dahmani-Mardas *et al.* (2010)
11	Common Bean	cv. BAT 93	EMS	Porch *et al.* (2009)
12	Radish	cv. Comet	EMS	Kohzuma *et al.* (2017)
13	Rape	cv. Express	EMS	Braatz *et al.* (2018)
14		cv. Express 617	EMS	Shah *et al.* (2018)
15		cv. R-o-18	EMS	Stephenson *et al.* (2010)

A shortcoming in TILLING is that only a portion of the target gene can be screened. Moreover, designating the most appropriate window is impeded by the restrained knowledge of the gene sequence information of the target plant species. For designing the suitable primer, the availability of the gene sequence is mandatory. An available gene sequence makes the target gene identification simpler. If sequence information is unavailable for the target crop species, the choice of gene window may account to the sequence information from a near relative crop or the gene model (Slade *et al.*, 2011).

The process of designing suitable primers is facilitated by software tools that work based on the information database on the homology to predict the mutational effect on protein function. **CODDLE** (Codons Optimized to Detect Deleterious Lesions), is one such a free web-based software, commonly used to design the primers for target gene identification. The function of CODDLE is to identify the regions where point mutations are likely to create a deleterious effect on genes and exploit the available information on the typology of the induced mutation, with the help of genomic sequence and corresponding coding sequence information available in its database. It also predicts the probability of missense and nonsense mutation. Ultimately it identifies the gene portion that is most desirable for TILLING and the same will be used for primer design. CODDLE also generates tabulated information on the selected amplicon and its possible effect of polymorphism induced by the mutagen (Bovina *et al.*, 2011).

Molecular Screening

It should be borne in mind that the molecular or genotypic screening for mutation should be executed on M_2 plants but not M_1 plants. This is why because the M_1 population contains the chimerism, which arises when multicellular tissues of the seed come in contact with the mutagen. The mutations induced in the germline cells are exclusively inheritable. The M_2 population, being the first non-chimeric population is a good subject for molecular screening for mutations.

Similarly, the phenotypic analysis, which reveals the gene of interest, is to be subjected to the M_3 individuals, since the M_2 population is a segregating population with wide variation. Once the M_2 individual is found confined with a mutation in the gene of interest, then the M_3 population, forwarded from M_2, should be ensured with homozygosity (since most mutations are recessive) and free from background mutations. The background mutations refer to the additional unwanted mutations in other genes that are likely to influence the desired mutation and they can be eliminated by crossing the mutant plants with the parent line. Beneficial is to create a database to store the information on mutated generations, phenotypes of M_2 and M_3 lines and the gene sequence and type of mutation induced.

The Pooling of the DNAs from Multiple M_2 Mutant Individuals

The DNA samples isolated from each M_2 plant are to be pooled before the genotypic screening, for obtaining maximum throughput of mutation discovery. The DNA samples are isolated from leaves at appropriate growth stages and the M_2 DNA library is constructed to identify the plants containing mutation within the gene of interest. It becomes necessary to pool the maximum number of individuals as possible to acquire enough number of mutant alleles. Moreover, intensifying the pooling level reduces the number of analyses required to screen the entire collection of mutants (Bovina *et al.*, 2011). The pooled DNA samples are bulked, arrayed on microtitre plates, and follows the PCR amplification of the DNA fragments of interest.

Heteroduplex Formation and Enzymatic Cleavage

The DNA pools are then submitted for window-specific amplication by PCR, along with properly designed primers. The amplified products are denatured and annealed, by heating and cooling. If the point mutation is found present in the amplified region, the mutant strand will reanneal with the wild type, showing a mismatch at the site of mutation in the fraction of the DNA pool. For high throughput TILLING purposes, an efficient heteroduplex detection method based on endonucleotypic cleavage and sizing on electrophoretic gel has been developed. An enzyme nuclease, a member of the S1 family, purified from celery (*Cel1*), cleaves the mismatches on the wither strand on the 3′-side and incubating sample pools in presence of S1 nuclease will generate short fragments.

Detection of Cleaved Fragments

Initially, the detection system was based on denaturing High-Pressure Liquid Chromatography (dHPLC), wherein the solid phase contains a differential affinity for single and double-stranded DNA, based on their different hydrophobicity. Having the column temperature just below the melting point of homoduplex, causes the mismatched fragment to melt out, as the duplex passes through the column, decreasing its hydrophobicity and therefore its retention time in the column. This method is pretty fast and highly specific.

Later the detection method leveled up into a more efficient system that works based on sizing by electrophoretic gels. This approach was implemented for LI-COR sequencers (LI-COR Inc., 2006; www.licor.com) that comprised of a denaturing polyacrylamide slab gel analyzer. This instrument visualizes the fragments based on the fluorescence detection, where the PCR amplification of the target fragments is executed in the presence of infra-red dye-labeled primers. The forward primer is usually 5′-end-labeled with a fluorescent dye IRDye700 (detected at ca. 700 nm) and the reverse primer is labeled with IRDye800 (detected at ca. 800 nm) (Bovina *et al.*, 2011).

For every round, the primers fluorescence emissions are detected by LI-COR and two electronic image file is generated of data from 700 nm and 800 nm channels. Regardless of the positive result obtained in the first round, the result must be validated to confirm the mutation. Once the mutation is discovered, the probability of the protein function is to be found. Software functions based on protein homology information are used to predict the damaging lesions. PARSESNP (Project Aligned Related Sequence and Evaluate SNPs; http://www.proweb.org/parsesnp/) is a web-based tool available for polymorphism analysis. A similar bioinformatics tool for TILLING is SIFT (Sorting Intolerant From Tolerant; http://blocks.fhcrc.org/sift/SIFT), which predicts whether an amino acid substitution affects the protein function based on sequence homology and physical properties of amino acids.[4]

Recovery of Mutants

A phenotypic analysis on mutated allelic series is an essential element for TILLING. Aforesaid chemical mutagenesis induces a large number of point mutations, in addition to those at the target locus, which are called as background mutations. They can be cleaned up to recover the mutant by backcrossing with the parent material ((Bovina *et al.*, 2011; Szarejko *et al.*, 2011). When the phenotype has been observed, additional steps are needed to prove that the phenotypic characteristics are controlled by the gene of interest. The protocol of TILLING is illustrated in Figure 12.1.

Recent Advances in Vegetable Improvement through TILLING: Case Studies

The TILLING technique is applied to identify genes favoring agronomic, physiological and quality improvement traits predominantly in field crops *viz.*, rice, wheat, maize and barley and those on vegetable crops are quite rare. Thus, an effort is extended here to summarize some significant TILLING based research works and the novel genes identified in vegetable crops.

Improved Shelf Life in Musk Melon

Musk Melon (*Cucumis melo* L.), following climacteric nature in its ripening behavior, is highly influenced by the plant hormone ethylene. The ethylene driven fruit softening is the key phenomenon in the process of fruit ripening and it gets

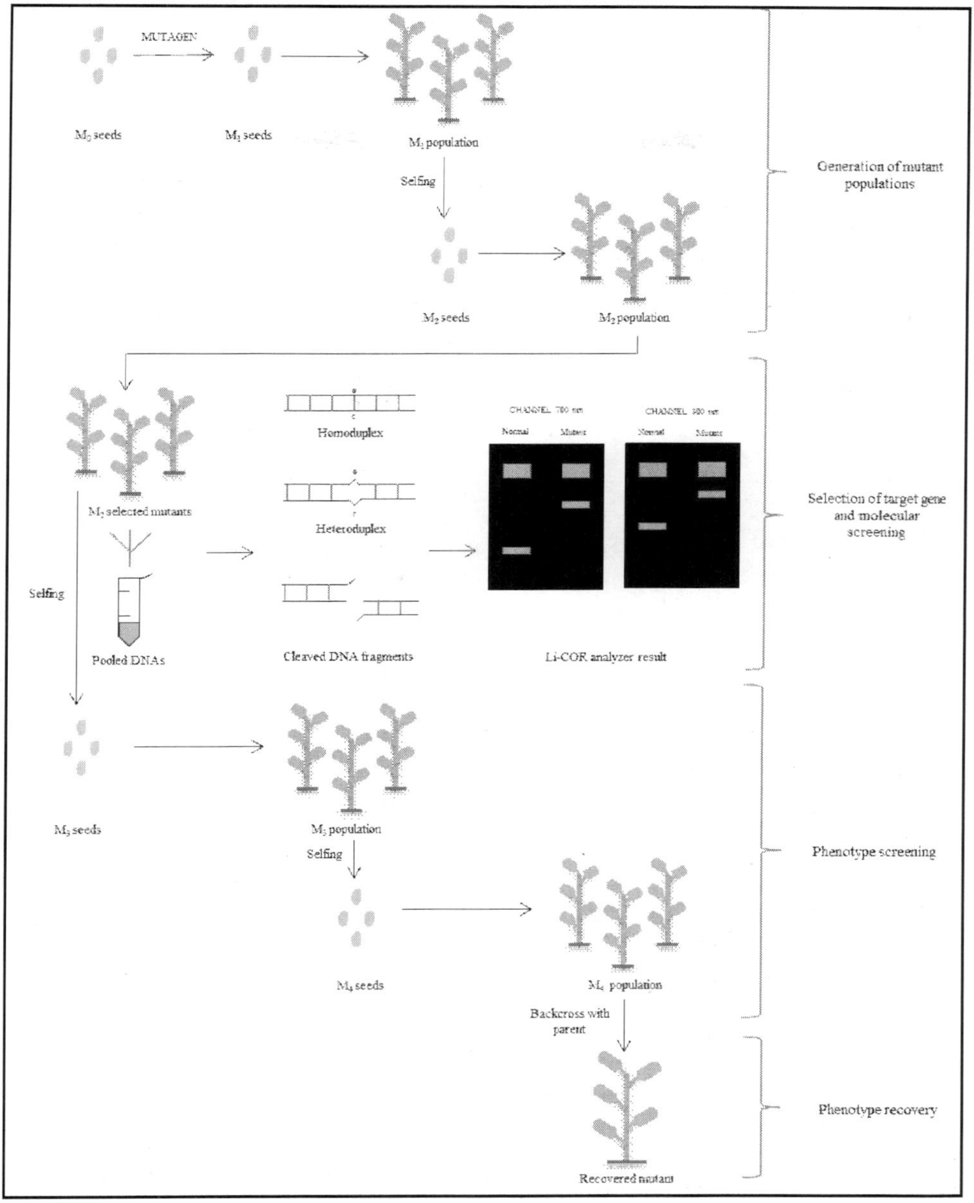

Figure 12.1: A Simple Illustration of TILLING Protocol.

hastened in a climacteric fruit during ripening, leading to excessive softening of fruit tissues, deterioration, and reduced shelf life. Thus, it becomes necessary to control over-ripening mediated by ethylene. Silencing of enzymes/regulators of ethylene biosynthesis pathways have been reported to inhibit fruit ripening and extended storage in tomato and melons (Ayub *et al.*, 1996). Besides, the ethylene biosynthesis pathway has been characterized and genes underlying are identified in *Arabidopsis.*

To engineer melon fruits with longer shelf life, a translational work had been taken up and a TILLING platform was set up with EMS as a mutagen. The mutations for the enzyme **CmACO1** (ACC Oxidase) that catalyze the final step of ethylene biosynthesis were screened. ACC Oxidase belongs to the multigene family and highly expressed during fruit ripening in climacteric fruits.

The screening results revealed the induction of two missense mutations – *L123F* and *G194D* of ACC Oxidase. The mutation *G194D* has taken place in a highly conserved position that is the potential to affect the enzymatic activity and exhibited delayed ripening coupled with improved shelf life while the mutation *L124F* was found to be a conservative mutation occurred away from the site encoding CmACO1 activity having no effect on enzyme activity.

TILLING of the gene pools discovered a mutation at the conserved G^{194} residue located at the active site of the enzyme. The mutant G194D was noticed with a longer fruit maturation period, enhanced firmness and a delay in the change of fruit color. The result pointed out that the glycine at position 194 encodes fruit maturation and shelf life via controlling the enzyme ACC Oxidase.[9] This study has yielded a new allele for improved shelf life in musk melon.

Extended Shelf-Life in Tomato

The tomato fruit ripening is accelerated by the softening of the cell wall of the fruit tissues. Expansin is a plant protein that has control in loosening the cell wall, cell enlargement, and other cell wall modification processes. During fruit ripening, the highly expressed expansin gene, **SlExp1** governs the fruit softening. An investigation aiming to discover new renditions on fruit softening, set up a TILLING population with EMS, and carried out screening for mutants impaired in the SlExp1 gene. Totally nine induced point mutation in SlExp1 coding regions were obtained and for further detailed phenotyping two missense mutations *viz.*, Slexp1-6-_w211S and Slexp1-7_Q213Stop were selected.

The fruits of the mutants Slexp1-6- and Slexp1-7 were characterized as firmer than control and the fruit firmness increased at all ripening stages. Subsequently, the enhanced firmness extended the fruit shelf life, as well. It was interpreted that the mutation in the SlExp1 gene led to the loss of expansin activity with modified cell wall polysaccharide composition. In the mutant Slexp1-7, there was a significant difference for uronic acid, neutral sugars, total sugar content, moreover, the xyloglucan structures of hemicellulose were affected in fruit pericarp (Minoia *et al.*, 2016). This study offers a deep perception on how the loss of SlExp1 function improves fruit shelf life.

Virus Resistance in Tomato

The translation initiation factor-mediated virus resistance is a widespread mechanism in plants (Robaglia and Caranta, 2006). To instigate their role, a study was taken by Piron *et al.* (2010) in tomato. The TILLING platform was developed with EMS, genes responsible for the translation initiation factors **eIF4E** and **eIF4G** were cloned and a molecular screening ran for mutation with induced resistance to viruses. In this study, 83 alleles with mutations are found and only 7 among them showed a mutation in the translation initiation factor eIF4E. Among these 7 mutant lines, only one splicing mutant of translation initiation factor, Sl_eIF4E G1485A was identified and characterized with immunity two Potato Virus Y and Pepper Mottle Virus. This mutant was found to be affected in a pre-mRNA splicing of Sl_eIF4E, non-functional for cap-binding function and encodes for a truncated mRNA lacking exons 2 and 3.

The reason attributing to the resistance was undeciphered, however, it was thought due to the mutations in amino acids (Charon *et al.*, 2008; Kang *et al.*, 2005; Von *et al.*, 1974). Therefore, the results have demonstrated the effectiveness of TILLING as a powerful reverse genetics tool to improve crop species regarding virus resistance.

The First Report on SlLAX1 Gene for Normal Leaf Development in Tomato

Besides, to improve traits in crop species, TILLING also provides a platform to study the undiscovered function of genes on the basic plant growth, development and metabolisms.

The leaves are the primary plant organs that are inevitable for photosynthesis. An important factor controlling morphogenesis is the hormone auxins that is synthesized in young leaves and shoot apexes. The auxin biosynthesis is controlled by the **AUX1/LAX** and PIN gene families, as influx and exflux carriers, respectively (Kramer and Bennett, 2006). The role of AUX1/LAX is well studied and authenticated for the root development and the same for leaf morphogenesis is neglected and underestimated.

In this investigation, an EMS mutagenized TILLING population was generated from tomato cv. Micro-Tom. Six mutants of allelic curly leaf phenotype were isolated. A mutation in **the SlLAX1** gene (Solyc09g014380) was observed and it was characterized as a 'curl mutant'. This study sheds light on the essentiality of the SlLAX1 gene for normal leaf development, without which the leaves exhibited curling. The gene SlLAX1 governed the normal development by balanced adaxial and abaxial pavement cell growth in tomato (Pulungan *et al.*, 2018).

Advantages of TILLING Strategy

Though the conventional mutation breeding leads to high yielding varieties in various crops, only the mutants conveying a visible alteration in the phenotype are identifiable and suffer from a lack of techniques to estimate the mutation frequency. In this context, the TILLING provides a direct measure of induced mutations. In addition to the advantage of gene discovery, utilization of TILLING can also be extended to DNA polymorphism assessment and crop improvement. Some of its major advantages are given below.

a. A Simple Procedure

The technique comprised of traditional chemical mutagenesis and double-dye far-red fluorescent detection demands no complicated calibrations and sophisticated apparatuses, thereby enabling mutant pool screening at ease for investigating and identifying the mutations in the specific genes (Colbert *et al.*, 2001).

b. The High Sensitivity of the Detection System

The enzyme *Cel 1*, containing the single-strand specific nuclease, is potential to cleave almost all the single base-pair mismatches of the heteroduplexes obtained from a variety of mutations (Minoia *et al.*, 2016; Oleykowski *et al.*, 1998).

c. High Efficiency in Mutant Identification

The high efficiency in the mutant identification is an interesting trait of TILLING featured by the high frequencies of mutagenesis and high-throughput screening capacity of the system (De-kai *et al.*, 2006). The consolidation of automatic sample-loading systems, high-throughput gel, prominent Li-COR analyzer, and salient image processing programs have transformed TILLING with a mighty screening capacity (Till *et al.*, 2003).

Conclusion

TILLING, as a powerful reverse genetics tool, has proved its efficiency in upgrading the chemical-induced mutagenesis. It provides greater scope for identifying mutations in the potential genes that are opined to be effete. It not only enables the generation of a new series of alleles for a phenotype of interest but also extends a new understanding of molecular and biochemical mechanisms of basic plant functions. Thus, it edifies the modern plant breeders to engineer crop plants with agronomic and quality considerations.

REFERENCES

Ayub R, Guis M, Ben Amor M, Gillot L, Roustan JP. 1996. Expression of ACC oxidase antisense gene inhibits ripening of cantaloupe melon fruits. *Nat Biotechnol.* 14: 862–866.

Baldet P, Bres C, Okabe Y, Mauxion J, Just D, Boumanoville C 2013. Investigating the role of vitamin C in tomato through TILLING identification of ascorbate-deficient tomato mutants. *Plant Biotechnology*. 30: 309-314. doi: 10.5511/plantbiotechnology.13.0622b.

Boualem A, Fleurier S, Troadec C, Audigier P, Kumar APK, Chatterjee M 2019. Development of a *Cucumis sativus* TILLinG platform for forward and reverse genetics. PLoS ONE. 9(5): e97963.

Bovina R, Talame V, Salvi S, Sanguineti MC, Tuberosa R. 2011. Discovery of chemically induced mutations by TILLING. In: Plant mutation breeding and biotechnology (Eds) Shu QY, Forster BP, Nakagawa H. Joint FAO/IAEA Division of Nuclear Techniques in Food and Agriculture, Vienna, Austria,; 257-268.

Braatz J, Harloff H, Emrani N, Elisha C, Heepe L, Gorb SN, 2018. The effect of INDEHISCENT point mutations on silique shatter resistance in oilseed rape

(*Brassica napus*). *Theoritical and Applied Genetics*, 131: 959-971. DOI: 10.1007/s00122-018-3051-4.

Charron C, Nicolai M, Gallois JL, Robaglia C, Moury B, 2008. Natural variation and functional analyses provide evidence for co-evolution between plant eIF4E and potyviral VPg. *Plant J*, 54: 56–68.

Chen C, Qing-zhi C, San-wen H, Shen-hao W, Xiao-hong L, Xiang-yang L, 2018. An EMS mutant library for cucumber. *Journal of Integrative Agriculture*. 17(7): 1612-1619.

Colbert T, Till BJ, Tompa R, Reynolds S, Steine MN, Yeung AT, McCallum CM, Comai L, Henikoff S. 2001, High-throughput screening for induced point mutations. *Plant Physiol*, 126(2): 480-484.

Dahmani-Mardas F, Troadec C, Boualem A, Leveque S, Alsadon AA, Alodss AA 2010 Engineering melon plants with improved fruit shelf life using the TILLING approach. *PLoS ONE*. 5(12): e15776.

De-kai W, Zong-Xiu, Yue-Zhi, 2006. Application of TILLING in plant improvement. *Acta Genetica Sinica*. 33(11): 957-964.

Gruszka D, Szarejko I, Maluszynski M. Sodium azide as a mutagen. 2012. In: Shu QY, Forster BP, Nakagawa H (Eds) Plant mutation breeding and biotechnology. CABI, the United Kingdom. 159–166.

Jones MO, Piron-Prunier F, Marcel F, Piednoir-Barbeau E, Alsadon AA, Wahb-Allah MA, 2012. Characterization of alleles of tomato light signaling genes generated by TILLING. *Phytochemistry*. 79: 78–86.

Just D, Garcia V, Fernandez L, Bres C, Mauxion J, Petit J, 2013. Micro-Tom mutants for functional analysis of target genes and discovery of new alleles in tomato. *Plant Biotechnology*. 30: 225–231.

Kang BC, Yeam I, Frantz JD, Murphy JF, Jahn MM. 2005. The pvr1 locus in Capsicum encodes a translation initiation factor eIF4E that interacts with Tobacco *etc*.h virus VPg. *Plant J*. 42: 392–405.

Kohzuma K, Chiba M, Nagano S, Anai T, Ueda MU, Oguchi R, 2017. Mutant selection in the self-incompatible plant radish (*Raphanus sativus* L. var. *sativus*) using two-step. *Breeding Science*. 67: 268–276.

Kramer, E.M. and Bennett, M.J. 2006. Auxin transport: a field in flux. *Trends Plant Sci*. 11: 382–386.

Kurowska M, Labocha-Paw³owska A, Gnizda D, Maluszynski M, Szarejko I. 2012. Molecular analysis of point mutations in a barley genome exposed to MNU and gamma rays. *Mutat. Res*. 738: 52–70.

Maluszynski M, Szarejko I, Bhatia R, Nichterlein K, Lagoda PJL 2009. Mutation techniques. In: Guimares E, Ceccarelli S, Weltzein E, Rajendran PG (Eds) Plant breeding book. FAO, Rome. 159–194.

Maluszynski M, Szarejko I, Maluszynska J. Mutation techniques. 2003. In: Thomas B, Murphy DJ, Murray BG (Eds) Encyclopedia of applied plant sciences. Elsevier Academic, the United States of America, 186–201.

Martin B, Ramiro M, Martý´nez-Zapater JM, Alonso-Blanco C. 2009, A high-density collection of EMS-induced mutations for TILLING in Landsberg erecta genetic background of Arabidopsis. *BMC Plant Biol.*, 9(147): 1–9.

McCallum CM., Comai L, Greene EA, Henikoff S. 2000. Targeted screening for induced mutations. *Nature Biotechnology*. 18: 455-457.

Minoia S, Boualem A, Marcel F, Troadec C, Quemener B, Cellini F, 2016. Induced mutations in tomato SlExp1 alter cell wall metabolism and delay fruit softening. *Plant Science*, 242: 195-202.

Oleykowski CA, Bronson Mullins CR, Godwin AK, Yeung AT. 1998, Mutation detection using a novel plant endonuclease. *Nucleic Acids Res*. 26(20): 4597-4602.

Piron F, Nicolai M, Minoia S, Piednoir E, Moretti A, Salgues A, 2010. An induced mutation in tomato eIF4E leads to immunity to two potyviruses. *PLoS ONE*. 5(6): e11313.

Porch TG, Blair MW, Lariguet P, Galeano C, Pankhurst C, Broughton WJ. 2009. Generation of a mutant population for TILLING common bean genotype BAT 93. *J. Amer. Soc. Hort. Sci*. 134(3): 348–355.

Pulungan SI, Yano R, Okabe Y, Ichino T, Kojima M, Takebayashi Y 2018. SlLAX1 is required for normal leaf development mediated by balanced adaxial and abaxial pavement cell growth in tomato. *Plant Cell Physiol*. 59(6): 1170-1186, DOI: 10.1093/pcp/pcy052.

Robaglia C, Caranta C. 2006.Translation initiation factors: a weak link in plant RNA virus infection. *Trends Plant Sci.*, 11: 40–45.

Shah S, Karunarathna L, Jung C, Emrani N. 2018. An APETALA1 ortholog affects plant architecture and seed yield component in oilseed rape (*Brassica napus* L.). *BMC Plant Biology*, 18: 380. doi: 10.1186/s12870-018-1606-9

Silletti MF, Petrozza A, Stigliani AL, Giorio G, Cellini F, D'Ambrosio C, 2013. An increase of lycopene content in tomato fruit is associated with a novel Cyc-B allele isolated through TILLING technology. *Mol Breeding*. 31: 665–674. DOI: 10.1007/s11032-012-9824-6.

Slade AJ, Moehs CP. 2011. Uses of TILLING for crop improvement. In: Plant mutation breeding and biotechnology (Eds) Shu QY, Forster BP, Nakagawa H. Joint FAO/IAEA Division of Nuclear Techniques in Food and Agriculture, Vienna, Austria, 359-369.

Stephenson P, Baker D, Girin T, Perez A, Amoah S, King GJ, *2010l*. A rich TILLING resource for studying gene function in *Brassica rapa. BMC Plant Biology. 10: 62.*

Szarejko I, Szurman-Zubrzycka M, Nawrot M, Marzec M, Gruszka D, kurowska M 2011. Creation of a TILLING population in Barley after Chemical mutagenesis with Sodium Azide and MNU. In: Biotechnologies for plant mutation breeding (Eds) Jankowicz-Cieslak J, Tai TH, Kumlejn J, Till BJ. Springer Open, Switzerland. 91-111.

Till BJ, Burtner C, Comai L, Henikoff S. 2004,Mismatch cleavage by single-strand specific nucleases. *Nucleic Acids Res.* 32(8): 2632-2641.

Till BJ, Reynolds SH, Greene EA, Codomo CA, Enns LC, Johnson JE, Burtner C, Odden AR, Young K, Taylor NE, Henikoff JG, Comai L, Henikoff S. 2003, Large-scale discovery of induced point mutations with high-throughput TILLING. *Genome Res.* 13: 524–530.

Von WD, Kahn A, Nielsen OF., 1974. Genetic regulation of chlorophyll synthesis analyzed with mutants in barley. *Science*. 184: 800-802.

Warren JJ, Forsberg LJ, Bees LS. 2003. The structural basis for the mutagenicity of O6-methylguanine lesion. *PNAS*. 103: 19701–19706.

Yeam I, Cavatorta JR, Ripoll DR, Kang BC, Jahn MM. 2007. Functional dissection of naturally occurring amino acid substitutions in eIF4E that confers recessive potyvirus resistance in plants. *Plant Cell.,* 19: 2913–2928.

Chapter 13

Mushrooms: A Potential Tool for Myco-remediation

Rishu Sharma

Department of Plant Pathology,
BCKV, Mohanpur, Distt. Nadia – 741 252, West Bengal

Mushrooms have been consumed since ancient times due to their high protein, flavor and low calorie attribute. Mushrooms have emerged as a vital bio-remediation tool, due to its potential to remediate various pollutants. Mycoremediation is based on the production of efficient and diverse enzymes, produced by mushrooms, for the degradation of various types of substrates and pollutants. Though, mushrooms are consumed as an edible product but have a pivotal role in waste degradation too. However, sometimes due to the biosorption process, mushroom fruit bodies tend to absorb the toxicants through their mycelium which makes them toxic and making them non-palatable. This chapter revolves around achievement and current status of mycoremediation technology based on the mushroom cultivation and its role in remediation of waste.

Mycoremediation, sometimes referred to as fungi remediation or mushroom remediation, is a form of bioremediation that uses fungi instead of bacteria to break down waste. Mycoremediation may offer a more sustainable and effective way to decompose waste than traditional bioremediation processes, yet is often unheard of, misunderstood and therefore rarely employed. Mushrooms have been integral part of many cuisines across the globe since time immemorial. The reference of mushrooms has been there in ancient scriptures like Vedas and Bible and their use as a source of many vitamins and minerals. Mushrooms have also been reported as nutraceuticals having anti-oxidant, anti-cancer, immunostimulatory, anti-inflammatory and anti-diabetic therapeutic properties (Barros *et al.*, 2007; Kim *et al.*, 2007; Synytsya *et al.*, 2009) (Table 13.1).

Table 13.1: Role of Mushrooms as an Important Product

Sl.No.	Mushroom	As a product	References
1.	*Pleurotus, Agaricus, Ganoderma, Schizophyllan commune, Grifola frondosa, Coriolus versicolor, Ganoderma lucidum*	Used as medicine to boost immune responses against cancer	Kodama *et al.* (2002); Gao *et al.* (2003)
2.	*Pleurotus, Agaricus*	Possess antimutagenic or antigenotoxic power to fight against cancer	Maehara *et al.* (2012)
3.	*Ganoderma lucidum, Phellinus rimosus, Pleurotus florida* and *Pleurotus pulmonaris*	Used as antioxidant and antitumor agent	Gameiro *et al.* (2013); Kang *et al.* (2012)
4.	*Pleurotus, Agaricus*	Used as food	Ajith and Janardhanan (2007)

Mycoremediation is the technology based on the natural biotransformation mechanisms of fungi to convert residues that might be considered as toxic for humans, into harmless molecules that may then be released to nature. From this perspective, it seems logical to assume that directed screening programs would and microorganisms with the ability to decompose a large variety of chemical structures. Fungi are able to produce a wide variety of enzymes and secondary metabolites as a result of changing environmental conditions. Both the nature of the substrate and fungal plasticity can then be combined to transform contaminated soils and restore devastated areas.

Fungi have been found to degrade pollutants from the environment, thus causing the myco-remediation. Fungus has the inherent capacity to break molecules, multi chain toxins into simpler and less toxic compounds. Fungus can remove heavy metals by channeling them into fruitbodies. The decontamination of soil and water from pollutants using microorganisms (bioremediators) is known asbioremediation (Rhodes, 2013). There are essentially two approaches, described as in situ and ex situ. In situ methods are those in which the contaminated material is treated onsite, whereas when the material is physically removed to be treated elsewhere it is referred to as ex situ. Methods of bioremediation offer means to degrade toxic organic materials, *e.g.* from oil spills, pesticides, and industrial waste, at the molecular level, converting them to more innocuous compounds. The ultimate goal of bioremediation is the full mineralization of contaminants, *i.e.* their transformation to CO_2, H_2O, N_2, HCl, *etc.* Heavy metal and radioactive cations, of course, cannot be decomposed but can be rendered into forms of low solubility, *e.g.* by a change in oxidation state, such as U (IV) (in UO_2) (Singh *et al.*, 2014), so that they remain less harmfully in the ground, or might be physically removed by phytoremediation or mycoremediation, which involves harvesting the plant or fungus.

White rot fungi are excellent myco-mediators of toxins held together by hydrogen carbon bonds. Enzymes secreted by white rotters include lignin peroxidases, manganes peroxidases and laccases. Extra-cellular lignin modifying enzymes have very low substrate specificity so they are able to mineraliaze a wide range of highly recalcitrant organo pullutants that are similar in structure to lignin.

The enzymes of the lignin degradation system of white rot fungi are extracellular, thus negating need to internalize the pollutants and enabling the fungi concentration of pollutants. White rot-fungi cannot utilize lignin as a source of energy for growth and instead require co-substrates as a carbon source. White rot fungi grow by hyphal extension and thus can reach pollutants in the soil in ways that other organisms cannot. White rot fungi, Lentinula edodes have been found to remediate the non-steroidal, anti-inflammatory drug piroxicam (Muszynska *et al.*, 2019)

Is Bioremediation the Future of Ecosystem Restoration?

Ecological Role of Mycelium: Mycelium contribute to biodiversity in nature by providing a nutrient source for microbes aiding in the decomposing process when there is an imbalance in the environment caused by the presence of a pollutant. Mycoremediated areas also provide the framework to jumpstart the ecological restoration process. While selecting mushrooms for remediating a toxic site choose species that naturally grow first. Spent mushroom compost is much more bio diverse and thus more effective than a pure culture in remediating a polluted site. The global mushroom industry has expanded very rapidly in the last two decades by the addition of newer types of mushrooms for commercial cultivation.

Mycelium is the vegetative heart of fungus which mushrooms can help save the world. It creates a network in soils for nutrient exchange among plants (Stamets, 2005). A fungi life cycle includes many stages, including hyphae that extend underground. Stamets believes, "that mycelium is the neurological network of nature."Fungus serves as nature's recycling agent. It typically secretes digestive enzymes into its environment that break down molecules for use as a food source.

Bioaugmentation

It is a category of bio-remediation. It involves introducing organisms that can degrade contaminants. Mycelium has been degrading molecules for hundreds of millions of years, so employing it for bioaugmentation is a natural strategy. Pollution is an unfortunate consequence of the technological advances made by mankind, and the methods to reverse contamination are often expensive and inefficient. The earliest well-known use of fungi in bioaugmentation came in 1985, when J.A. Bumpus and his team used a fungus called Phanerochaete chrysosporium to degrade dichlorodiphenyltrichloroethane (DDT), a popular pesticide that proved to be harmful to several aspects of the environment. Bumpus and his team also showed that P. chrysosporium degraded other organic pollutants. The use of DDT has been banned for many years, but other pollutants are more ubiquitous than ever.

Polyaromatic Hydrocarbons

Two PAH-contaminated soils used two types of white rot fungal cultures to degrade twelve PAHs, and examined the effects that mycoremediation had on the toxicity of the soils. Seed germination tests were also performed before and after the mycoremediation. It showed that germination rates for mustard seeds were related to soil contaminant concentrations, with higher concentrations being more limiting to growth. After fourteen weeks of incubation, the fungus Irpex lacteus removed 67 per cent of the PAH fluorine, and 57 per cent of fluoranthene among others.

Also, it was observed that mycoremediation of PAH contaminated soil is more complex when practiced in situ. When mobilizing agents are introduced in a laboratory setting, the ability of mycelium to degrade compounds can be less effective. This is the result of lower availability of the PAH in the soil. Another approach to in situ contaminated soil occurred in 2007 on a condemned industrial site in Italy. Heavy metals were first leached out with citric acid, and then the soil was treated with fungal mycelium of the genus Allescheriella to degrade the remaining organic pollutants. This in situ study resulted in the remediation of 63.8 per cent of organic pollutants. Integrated techniques such as this could prove most useful when contaminants are less available for remediation.

DEPH

Some pollutants are more subtle than pesticides and byproducts of combustion. Di(2-ethylhexyl)phthalate (DEHP) is a common additive to increase strength and flexibility in plastic medical devices, namely polyvinylchloride (PVC) blood storage bags. Because the DEHP is not chemically bonded to PVC, it washes away and ends up in watersheds when the plastics are buried. DEHP has been shown to be harmful to animals and humans, especially in early stages of development. In 2012, scientists in India showed that DEHP still bonded to PVC blood bags could be remediated with three types of mycelial fungi, either acting separately, but most effectively when used together. Over 99 per cent of the toxin was remediated, leaving only PVC that could be recycled for future use. The mycoremediation used in this case was very efficient, cost-effective, and eco-friendly.

Bioaugmentation has been studied for several years to combat another common, more obvious pollutant: trinitrotoluene, or TNT. In 1999, the fungus Bjerkandera adusta was shown to degrade TNT, but the in situ applications had limits. For a fungus to effectively degrade TNT in soil in situ to be most effective and thought to have large scale practical use in situ. This fungus degraded 80 per cent of the TNT in only 2.5 months.

Mycoremediation has provided yet another example of pollutant treatment. Stamets and his team were given four piles of soil, contaminated at 20,000 parts per million (ppm) petroleum hydrocarbons. One pile was the untreated control, one was treated with enzymes, one with bacteria, and the last with the mycelium of the familiar oyster mushroom. After four weeks, three of the piles were still black and lifeless, while the mycelium treated pile had become overtaken with oyster mushrooms. Over the next few weeks, the mushrooms produced spores and died. In doing so, they attracted insects hydrocarbons for the mycelium treated pile was less than 200 ppm. This example shows the remarkable effects of mycoremediation.

Myco-remediation is the Bioremediation with Fungi

Some of the prospects of using fungi, principally white-rot fungi, for cleaning contaminated land are surveyed. The white rot fungi are so effective in degrading a wide range of organic molecules is due to their release of extra-cellular lignin modifying enzymes, with a low substrate-specificity, so they can act upon various molecules that are broadly similar to lignin. The enzymes present in the system employed for degrading lignin include lignin-peroxidase (LiP), manganese peroxidase (MnP), various H_2O_2 producing enzymes and laccase. The degradation can be augmented by adding carbon sources such as sawdust, straw and corn cob at polluted sites.

Fungi feature among Nature's most vigorous agents for the decomposition of waste matter, and are an essential component of the soil food web (Rhodes, 2012), providing nourishment for the other biota that live in the soil. The forest floor is covered with leaf litter from the previous season, which plants cannot use directly to grow on, because the fallen leaves are too tough to be broken down and digested; thus, any nutrients they contain are locked within them. The key organism for breaking down the leaf litter is fungus: strictly, mycelium – the vegetative part of the fungus – which we often observe as fine, white threads that grow out from dead wood, leaves, *etc.* Indeed, fungi are the only organisms on Earth that can decompose wood. The mycelium exudes powerful extracellular enzymes and acids that are able to decompose lignin and cellulose, the two essential components of plant fibre. As the fungus breaks down wood and leaves, a rich material called humus is formed In the natural ecosystem, a realm of organisms from different kingdoms make their assault on those different substrates that are present, and the rate of degradation becomes maximal when there is a good supply of nutrients in the soil, *e.g.* N, P, K and other essential inorganic elements (Rhodes, 2013). Aspergillus and other moulds are highly efficient in decomposing starches, hemicelluloses, celluloses, pectins and other sugar polymers, and some aspergilli can degrade such intractable substrates as fats, oils, chitin, and keratin. To achieve a successful mycoremediation, the correct fungal species must be selected to target a particular pollutant, for which a simple screening procedure has been described (Matsubara *et al.*, 2006). An encyclopaedic overview of the research literature on the action of fungi on organic pollutants, up to 2006 is available (Singh, 2006), along with the beautifully illustrated "Mycelium running: how mushrooms can help save the world" (Stamets, 2005), which serves to provide a hands-on guide to growing fungi and applying them to remediating contamination, *e.g.* from oil spills and chemical toxins, on the practical scale.

Uptake of Heavy Metals using Myco-remediation

Bioremediation is a way of cleaning up heavy metals using biomass (or microorganisms) through the processes of biodegradation, biosorption, bioaccumulation and bioconversion operating in different ways (Kulshreshtha *et al.*, 2014; Mosa *et al.*, 2016). The microorganisms bind to heavy metals and concentrate them (Joutey *et al.*, 2015). Mushrooms can be employed for decontamination of the polluted environment. Mushrooms can build up heavy metals in high concentrations in their bodies above maximum permissible concentrations and can act as an effective biosorption tool (Das 2005). High accumulation potential and shorter life span are some of the advantages of using mushrooms as biosorbents. Mushrooms belonging to the genera including Agaricus, Boletus, Armillaria, Polyporus, Russula, Pleurotus, Termitomyces have been investigated by some researchers for the uptake of heavy metals (Raj *et al.*, 2011). Bio based syrgies have been explored due to indiscriminate use of fungicides, pesticides and effluents from various factories has led to severe contamination of environmental segments by heavy metals. Heavy metals are non-biodegradable and tend to be biomagnified in the food chain (Singh *et al.*, 2008). They pose a risk to human health when transferred via food chain and can further lead to toxic effects in organisms even in trace amounts. Conventional methods like chemical precipitation, adsorption, ion exchange, reverse osmosis and electrodialysis, to get rid of heavy metal burden of the environment, have their own shortcomings. These methods offer limitations like slow metal precipitation and incomplete removal, generation of contaminated sludge requiring careful disposal, high cost involved in the processes (Aziz *et al.*, 2015,Gunatilake 2015, Ayangbenro 2017, Firdousi 2017, Ahalya *et al.*, 2003), high energy and reagent requirements and clogging of membranes. In this scenario, it is important to opt for an economically feasible and effective treatment method which is free from these limitations and is able to translate the need of removal of heavy metals in terms of eco-friendly approach. These functional characteristics are mainly due to their chemical composition. Apart from this, Biosorption is a passive process and heavy metals get adsorbed on the surface of the biosorbent (Velásquez and Dussan 2009) exhibiting the tolerance of biosorbent towards heavy metals. The mechanisms like extracellular (chelation and cell wall binding) and intracellular (binding to compounds like proteins) sequestration of heavy metals have been proposed as mechanisms for heavy metals tolerance in fungi (Fawzy *et al.*, 2017). Biosorbents from mushrooms can be prepared from mycelium or fruit body (live or dead) and spent mushroom substrate (SMS). The factors like the presence of microbial population, the availability of contaminants to these organisms, metal ion concentration and environmental factors like temperature, pH and the presence of nutrients affect the biosorption process in totality (Prakash 2017). The process includes precipitation, ion exchange, electrostatic interaction, the redox process, *etc.* (Yang *et al.*, 2015). The biological process of remediation display features like economic viability (Ayangbenro 2017) and repeated use of biomass, selective metal binding and effective desorption and recycling of desorbents. Different microorganisms like algae, bacteria, fungi, yeast have been employed to carry out biosorption. The potential of fungal biomass as biosorbent has been accepted for the removal of heavy metals and radionuclides from polluted waters because of their excellent metal binding properties and tolerance towards metals and adverse environment like diverse pH and temperature conditions (Qazilbash 2004; Anand *et al.*, 2006; Yazdani *et al.*, 2010; Salman *et al.*, 2014). Fungi have been reported to exhibit the ability to chemically modify or affect their bioavailability (Prakash 2017).

Uptake and bioaccumulation studies have been done on Pleurotus species grown on metal-enriched substrates (Jain *et al.*, 1988, 1989). Pleurotus sajor-caju, grown on metal-enriched substrate duckweed, has been found to accumulate Cd content above permissible limits (Jain *et al.*, 1988). The bioaccumulation potential of Hg by P. ostreatus grown on artificial compost has also been studied by Bressa *et al.* (1988). They have the ability to enhance the nutritional content of the soil found in these areas (Adenipekun 2008) and bioremediate (Radulescu *et al.*, 2010). Subsequent to uptake, the metals are distributed unevenly within the fruiting bodies of mushrooms. The highest concentrations have been observed in the spore-forming part followed by rest of the cap and stipe (Gabriel *et al.*, 1996). Cd has been found to be present in higher concentrations in caps (22–56 mg/kg dry wt) than in stipe

(13–36 mg/kg dry wt) (Favero *et al.*, 1990a) in P. ostreatus. Pleurotus species have been found to show resistance to high Cd concentrations (Gabriel *et al.*, 1996). Their capacity to accumulate the heavy metals can lead to their immobilization but ingestion by other organisms can result in transfer along food chain (Osman and Bandyopadhyay 1999).The amounts of Pb, As, Fe, Cd and Hg in P. ostreatus available in the market (Accra, Ghana) have been found to be unsafe (Quarcoo and Adotey 2013).

A number of factors affect the uptake and biosorption of heavy metals. The biosorption by the target species varies with the type of metal, its concentration and composition of substrate (Javaid and Bajwa 2008).In the biosorption study conducted by Adhikari *et al.* (2004), P. florida has been found to sorb heavy metals in the order of Cd > Cr and to accumulate 1.2–2.5 per cent more Cd than Fusarium oxysporum, Penicillium species and Aspergillus awamorii. P. ostreatus have carboxylic, amino, thiol, phosphate and hydroxide groups on the cell wall helping in the biosorption of heavy metals (Banerjee and Nayak 2007; Javaid *et al.*, 2011). Boamponsem *et al.* (2013) reported that the age of the fungal fruiting body or its size is of less importance in the accumulation of heavy metals by mushrooms. Fruiting bodies of P. ostreatus immobilised in calcium alginate were shown to be effective in removing Pb and Co from solution (Xiangliang *et al.*, 2005, 2009) (Table 13.2–13.4).

Table 13.2: Use of Mushrooms in the Degradation of Pollutants

Fungal Strain	*Waste/Pollutant*	*Remarks*	*Reference*
Pleurotus ostreatus	Oxo biodegradable plastics	Mushroom degraded plastic and grew on it	Da luz *et al.*, 2013
Lentinula edodes	2,4 dichlorophenol	Mushroom degraded 2,4 dichlorophenol by using vanillin as activator	Tsujiyama *et al.*, 2013
Pleurotus pulmonarius	Radioactive cellulosic base waste	Waste containing mushroom mycelia was solidified with Portland cement and used as first barrier against the release of radio contaminants	Eskander *et al.*, 2012

Table 13.3: Role of Mushroom in Degradation of Pollutants

Sl.No.	*Mushroom spp.*	*Waste/Pollutants*	*Remarks*	*References*
1	*Pleurotus ostreatus*	Oxo-Biodegradable plastic	Mushrooms degraded the plastic and grew on it.	da Luz *et al.* (2013)
2	*Lentinula edodes*	2,4-dichlorophenol	Mushrooms degraded 2,4-dichlorophenol (DCP) by using vanillin as an activator	Tsujiyama*et al.* (2013)
3	*Pleurotus pulmonarius*	Radioactive cellulosic based waste	Waste containing mushroom mycellium was solidified with portland cement and then this solidified waste act as first barrier against the release of radiocontaminants	Eskander*et al.* (2012)

Sl.No.	Mushroom spp.	Waste/Pollutants	Remarks	References
4	*Jelly* sp., *Schizophyllum commune* and *Polyporous* sp.	Malachite green	99.75 per cent (*Jelly* sp.), 97.5 per cent (*Schizophyllum commune*), 68.5 per cent (*Polyporous sp.*2) dye was degraded in 10 days	Rajput *et al.* (2011)
5	*Pleurotus pulmonarius*	Crude oil	Crude oil was degraded	Olusola and Anslem(2010)
6	Coriolus versicolorMKACC 52492	PAH	Mushroom possesses ability to degrade Poly-R 478 which decides its suitability to degrade PAH. Lignin-modifying enzymes laccase, manganese-dependent peroxidase (MnP), and lignin peroxidase (LiP)was found to produce for degradation	Jang *et al.* (2009)

Table 13.4: Removal of Pollutants by Biomass of Mushroom using Biosorption Process

Sl.No.	Mushrooms spp.	Pollutants	Remarks	References
1	*Agaricus bisporus, Lactarius piperatus*	Cadmium (II) ions	Wild *L.* piperatus showed higher removal efficiencyon Cd(II) ions compared to the cultivated *A. bisporus*	Nagy *et al.* (2013)
2	*Fomes fasciatus*	Copper (II)	Mushroom is efficient in biosorption of Cu (II) ions and hot-alkali treatment increased their affinity for Cu (II) ions	Sutherland and Venkobachar (2013)
3	*Pleurotus platypus, Agaricus bisporus, Calocybe indica*	Copper, Zinc, Iron,Cadmium, Lead, Nickle	Mushrooms are efficient biosorbent for the removal these ions from aqueous solution	Lamrood and Ralegankar (2013)
4	*Flammulina velutipes*	Copper	Mushroom compost used as biosorbent forremoving copper ions from aqueous solution	Luo *et al.* (2013)
5	*Pleurotus tuber-regium*	Heavy metals	Pleurotus tuber-regiumbiosorb the pollutant heavymetals from the soil artificially contaminatedwith some heavy metals	Oyetayo *et al.* (2012)
6	*Pleurotus ostreatus*	Cadmium	Mushroom possess biosorption capacity andmechanism of biosorption was observed	Tay *et al.* (2011)
7	*Pleurotus sajor-caju*	heavy metal Zn	Mushrooms biosorb the heavy metals	Jibran and Milsee Mol (2011)

Bioremediation using Fungi

White-rot fungi digest lignin by the secretion of enzymes and give a bleached appearance to wood, from undissolved cellulose, hence their name. In contrast, brown-rot fungi degrade cellulose, leaving lignin as a typically brownish deposit. These fungi also cause chequered, cubical cracking and shrinking in wood, which is frequently apparent on felled confer trees (Stamets, 2005). It has been estimated that some 30 per cent of the literature on fungal bioremediation is concerned with white-rot fungi (Singh, 2006). As we shall see, there are particular mechanisms implicit to white-rot over other kinds of fungi, which offer advantages, *e.g.* over the use of bacteria, as a means for bioremediation In particular, bacteria need to be pre-exposed to the particular pollutant they are intended to degrade, in order to induce those enzymes that are required to accomplish the task. There is, furthermore, a pollutant concentration level below which the enzymes are not expressed in bacteria, thus limiting the technology (Adenipekun and Lawal, 2012). A very large range of organic molecules are susceptible to the actions of various strains of white-rot fungi, to varying degrees, and even normally highly intractable and persistent substances, including polyaromatic hydrocarbons (PAH), may be degraded by them (Singh, 2006). The white rot fungus Phanerochaete chrysosporium is an ideal model for bioremediation by fungi, since it is more efficient than other fungi or microorganisms in degrading toxic or insoluble materials. It presents simultaneous oxidative and reductive mechanisms which permit its use in many different situations, both regarding the type of contamination, its degree, and the nature of the site itself. A number of other white-rot fungi also can degrade persistent xenobiotic compounds, *e.g.* Pleurotus ostreatus, Trametes versicolor, Bjerkandera adusta, Lentinula edodes, Irpex lacteus, Agaricus bisporus, Pleurotus tuberregium,Pleurotus pulmonarius. (Singh, 2006; Adenipekun and Lawal, 2012). Soils may also be decontaminated from crude oil, with the requirement that lignocellulosic substrates (*e.g.* sawdust straw and corn cob) must also be provided, to support the growth of fungal species in the soil (Lang *et al.*, 1995). Other toxic materials that have been successfully degraded using white-rot fungi are: polychlorinated biphenyls and dioxins, pesticides, phenols and chlorophenols, effluents from pulp and paper mills, dyestuffs and heavy metals (Singh, 2006). However, it has been shown that the degradation of Benzopyrene by Pleurotus ostreatus is strongly influenced by the presence of heavy metal cations and mediators such as vanillin and 2,22-azinobis-(3- ethylbenzothiazoline-6-sulfonate). A 15 mM concentration of copper was found to best enhance the degradation (74.2 per cent), which was progressively worsened as the Cu concentration increased. The extent of degradation was increased to 83.6 per cent when 5 mM of vanillin was included in the medium (Bhattacharya *et al.*, 2014). The possibility is offered, therefore, that the presence of vanillin (a breakdown product of lignin) might enhance the process of mycoremediation using white-rot fungi in actual field-applications. It has been demonstrated (Isikhuemhen *et al.*, 2011) that L. Squarrosulus can degrade cornstalks significantly after 30 days, with a maximum lignocellulolytic enzyme activity being achieved on day 6 of cultivation, to generate exopolysaccharides. Thus, L. squarrosulus might prove very effective in the industrial pretreatment and biodelignification of lignocellulosic biomass. The main reason that white-rot fungi are active to such a wide range of compounds is their release of extracellular lignin-modifying enzymes, with a low substrate specificity,so they can act upon various molecules that are broadly similar to lignin (Adenipekun and Lawal, 2012). The enzymes present in the system employed for degrading lignin include lignin-peroxidase (LiP), manganese peroxidase (MnP), various H_2O_2 producing enzymes (Kirk and Farrell, 1987) and laccase, although the three types of enzymatic activity are not present in all lignolytic fungi.

To use white-rot fungi successfully for bioremediation, knowledge must be taken from such fields as fungal physiology, biochemistry, enzymology, ecology, genetics, molecular biology and engineering among other cognate subjects. A four-phase strategy has been advocated (Lamar and White, 2001): bench-scale treatability, on-site pilot testing, production of inoculum, and finally full-scale application. Substrates such as wood chips, wheat straw, peat, corn cobs, sawdust, a nutrient-fortified mixture of grain and sawdust, bark, rice, annual plant stems and wood, fish oil, alfalfa, spent mushroom compost, sugarcane bagasse, coffee pulp, sugar beet pulp, okra, canola meal, cyclodextrins, and surfactants can be used in inoculum production both off-site or on-site, or as mixed with contaminated soils to improve the processes of degradation (Singh, 2006). It is critical to attain the correct nitrogen/carbon ratio in the substrates used, so to avoid any impeding effect on the efficiency of the fungi in the bioremediation process. Fungal inocula coated with alginate, gelatin, agarose, carrageenan, chitosan, *etc.*, in the form of pellets, may offer a better outcome than with inocula produced using bulk substrates. The latter approach, termed encapsulation, is derived from the mushroom spawn industry, and both preserve the viability of the inoculum and contributes nutrients to maximally support the degradation of pollutants. Fungal inoculum may also be obtained by solid state fermentation. Such inoculum preparation methods enhance the likelihood of success in the first phase (above), while good technical and engineering vitalise the second phase. Success in stages three and four depends on the exact remediation practices employed for the monitoring, optimisation, continuity and maintenance of the process overall. Native microbial populations will also provide a potential competition to the mycoremediation process, but there is, as yet, a lack of defined protocols to eliminate such influences. There are some patents available which refer to the subject of remediation using white-rot fungi (Singh, 2006) (Table 13.5).

Table 13.5: Role of Mushroom Species in Bioconversion of Waste

Sl.No.	Mushroom spp.	Bioconversion of Waste	Remarks	References
1	*Pleurotus citrinopileatus*	Handmade paper and cardboard industrial waste	Successfully cultivated. Basidiocarps possessed good nutrient content and no genotoxicity	Kulshreshtha *et al.* (2013)
2	*Pleurotus ostreatus*	Extract from the sawdust	Biomass of mushroom was produced in submerged liquid culture were analyzed	Akinyele *et al.* (2012)
3	*Volvariella volvacea*	Agro-industrial residues such as cassava, sugar beet pulp, wheat bran and apple pomase	Enzyme activities were measured during the fermentation of substrates	Akinyele *et al.* (2011)

Sl.No.	Mushroom spp.	Bioconversion of Waste	Remarks	References
4	*Pleurotus florida*	Handmade paper and cardboard industrial waste	Successfully cultivated. Basidiocarps possessed normal morphology and no genotoxicity	Kulshreshtha *et al.* (2010)
5	*Pleurotus*	Cotton waste, rice straw, cocoyam peels and sawdusts of *Mansonia altissima, Boscia angustifolia* and *Khayai vorensis*	Successfully cultivated with good crude protein, crude fat and carbohydrate contents in sporophores.	Kuforiji and Fasidi (2009)
6	*Pleurotus eous* and *Lentinus connotus*	Paddy straw, sorghum stalk, and banana pseudostem	Waste successfully bioconverted by mushroom with good biological efficiency	Rani *et al.* (2008)
7	*Pleurotus tuber-regium*	Nigerian trees; *Terminalia superba, Mansonia altissima, Holoptelia grandis* and *Milicia excelsa*	Grow on trees	Jonathan *et al.* (2008)
8	*Pleurotus tuber-regium*	Cotton waste, sawdust of *Khayai vorensis* and rice straw	Sclerotia propagated on groundnut shells and cocoyam peels with lipase and phenoloxidase; cellulase, carboxymethylcellulase enzymatic activities	Kuforiji and Fasidi (2008)
9	*Lentinula edodes*	Eucalyptus waste	Successfully convert this waste and qualitative and quantitative changes were also measured	Brienzo *et al.* (2007)
10	*Lentinula edodes*	Vineyard pruning (VP), barley straw (BS), and wheat straw	Bioconversion of VP waste with shortest primordium formation, highest biological efficiency, highest yield and shortest production cycle (6 days)	Gaitán-Hernández *et al.* (2006)
11	*Lentinula tigrinus*	Wheat straw	Characterize the production of lignocellulosic enzymes and bioconvert the wheat straw	Lechner and Papinutti (2006)
12	*V. volvacea*	Banana leaves (*Musa sapien tumlina*)	Efficient bioconversion with good yield	Belewu and Belewu (2005)

Conclusions and Future Thrust

Fungi has a great potential in decomposing in situ intractable, persistent, and highly toxic pollutants, including TNT(2,4,6-trinitrotoluene), DDT. Mushrooms being a food crop, its mycelia have a huge potential in myco-remediation along with spent mushroom substrate (Kapahi *et al.*, 2017). The mycelia of Pleurotus ostreatus have been found to be effective in converting 95 per cent PAH's present in diesel oil in duration of four weeks to non-toxic compounds. It has been observed that the naturally present community of microbes acts in concert with the fungi to decompose the contaminants, finally to CO2 plus H2O. Wood-degrading fungi are extremely effective in decomposing toxic aromatic pollutants from petroleum and also chlorinated persistent pesticides (Rhodes, 2013). Mycofiltration is a similar procedure, in which mycelia are used as a filter to remove toxic materials and microorganisms from water in the soil. It has been suggested that a "Mycological Response Teams" should be employed to recycle and rebuild the soil and edible mushrooms might be grown for the purposes of mycoremediation, and the prospects of whether they would be safe to eat afterwards are considered (Staments 2005, Kulshreshtha *et al.*, 2014). Naturally this depends on the exact nature of the pollutants and uptake of heavy metals, depending on their sorption capacity and causing toxicity in the mushroom fruit bodies while some organic soil contaminants might be decomposed without so imparting toxicity. Thus, we need to critically analyse the possibility of using mushrooms in bioremediation considering health hazards and exploring these studies to frame methodologies for in situ and ex situ bioremediation using mushrooms in the coming years.

REFERENCES

Adenipekun, CO and Lawal R (2012) Uses of mushrooms in bioremediation: A review. *Biotechnol. Molec. Biol. Rev.*, 7(3), 62-68.

Adenipekun CO (2008) Bioremediation of engine-oil polluted soil by Pleurotus tuber-regium Singer, a Nigerian white-rot fungus. *Afr J Biotechnol* 7: 55–58.

Agrahar-Murugkar D, Subbuakshmi G (2005) Nutritional value of edible wild mushrooms collected from the Khasi hills of Meghalaya. *Food Chem* 89: 599–603.

Ahalya N, Ramachandra TV, Kanamadi RD (2003) Biosorption of heavy metals. *Res J Chem Environ* 7(4): 71–79.

Ajith TA, Janardhanan KK (2007) Indian Medicinal Mushrooms as a Source of Antioxidant and Antitumor Agents. *J Clin Biochem Nutr* 40: 157–162, doi: 10.3164/jcbn.40.157.

Akinyele BJ, Olaniyi OO, Arotupin DJ (2011) Bioconversion of selected agricultural wastes and associated enzymes by Volvariellavolvacea: An edible mushroom. *Res J Microbiol* 6: 63–70, doi: 10.3923/jm.2011.63.70.

Akinyele JB, Fakoya S, Adetuyi CF (2012) Anti-growth factors associated with Pleurotus ostreatus in a submerged liquid fermentation. *Malaysian J Microbiol* 8: 135–140.

Akpaja EO, Nwogu NA, Odibo EA (2012) Effect of some heavy metals on the growth and development of Pleurotus tuber-regium. *Mycosphere* 3: 57–60.

Akyüz M, Kirbaŏ S (2010) Element contents of Pleurotuseryngii(DC. Ex Fr.) Quel. var. eryngii grown on some various agro-wastes. *Ekoloji* 19(74): 10–14.

Anand P, Isar J, Saran S, Saxena RK (2006) Bioaccumulation of copper by Trichodermaviride. *Bioresour Technol* 97: 1018–1025.

Arbanah M, MiradatulNajwa MR, Ku Halim KH (2012) Biosorption of Cr(III), Fe(II), Cu(II), Zn(II) ions from liquid laboratory chemical waste by *Pleurotus ostreatus. Int J Biotechnol Wellness Ind* 1: 152–162.

Arbanah M, MiradatulNajwa MR, Ku Halim KH (2013) Utilization of Pleurotus ostreatusin the removal of Cr(VI) from chemical laboratory waste. IntRefreed J EngSci 2(4): 29–39.

Arica MY, Arpa C, Kaya B (2003) Comparative biosorption of mercuric ions from aquatic systems by immobilized live and heat-inactivated Trametesversicolorand Pleurotussajur-caju.BioresourTechnol 89: 145–154.

Ayangbenro Babalola (2017) A new strategy for heavy metal polluted environments: a review of microbial biosorbents. Int J Environ Res Public Health 14: 94.

Aziz HA, Adlan MN, Ariffin KS (2015) Heavy metals (Cd, Pb, Zn, Ni, Cu and Cr(III)) removal from water in Malaysia: post treatment by high quality limestone. BioresourTechnol 99(6): 1578–1583.

Baldrian P, Gabriel J (2002) Copper and cadmium increase laccase activity in Pleurotus ostreatus. FEMS MicrobiolLett 206: 69–74.

Baldrian P, Gabriel J (2003) Lignocellulose degradation by Pleurotus ostreatus in the presence of cadmium. FEMS MicrobiolLett 220: 235–240.

Baldrian P, In Der Wiesche C, Gabriel J, Nerud F, Zadraz¡il F (2000) Influence of cadmium and mercury on activities of ligninolytic enzymes and degradation of polycyclic aromatic hydrocarbons by *Pleurotus ostreatus* in soil. Appl Environ Microbiol 66: 2471–2478.

Banerjee A, Nayak D (2007) Biosorption of no-carrier-added radio-nuclides by calcium alginate beads using 'tracer packet' technique. *Bioresour Technol* 98: 2771–2774.

Barcan VS, Kovnatsky EF, Smetannikova MS (1998) Absorption of heavy metals in wild berries and edible mushrooms in an area affected by smelter emissions. Water Air Soil Pollut 103: 173–195.

Barros L, Baptista P, Estevinho LM, Ferreira ICFR (2007) Bioactive properties of the medicinal mushroom *Leucopaxillus giganteusmycelium* obtained in the presence of different nitrogen sources. Food Chem 105: 179–186. doi: 10.1016/j.foodchem.2007.03.063.

Belewu MA, Belewu KY (2005) Cultivation of mushroom (Volvariella volvacea) onbanana leaves. African J Biotechnol 4: 1401–1403.

Bhattacharya S, Das A, Prashanthi K, Palaniswamy M, Angayarkanni J. 2014. Mycoremediation of Benzo[a]pyrene by Pleurotus ostreatus in the presence of heavy metals and mediators. *Biotech.* 4(2): 205-211. doi: 10.1007/s13205-013-0148-y.

Boamponsem GA, Obeng AK, Osei-Kwateng M, Badu AO (2013) Accumulation of heavy metals by Pleurotus ostreatusfrom soils of metal scrap sites.Int J Curr Res Rev 5(4): 01–09.

Bressa G, Coma L, Costa P (1988) Bioaccumulation of Hg in the mushroom Pleurotus ostreatus. Ecotoxicol Environ Safe 16: 85–89.

Brienzo M, Silva EM, Milagres AM (2007) Degradation of eucalyptus waste components by Lentinulaedodes strains detected by chemical and near-infrared spectroscopy methods. Appl BiochemBiotechnol 141: 37–50.

Brunnert H, Zadraz/il F (1983) The translocation of mercury and cadmium into the fruiting bodies of six higher fungi. A comparative study on species specificity in five lignocellulolytic fungi and the cultivated mushroom *Agaricus bosporus. Eur J Appl Micorbiol Biotechnol* 17: 358–364.

Carol D, Kingsley SJ, Vincent S (2012) Hexavalent chromium removal from aqueous solutions by Pleurotus ostreatusspent biomass. Int J Eng Sci Technol 4(1): 7–22.

Christopher J. Rhodes (2014) Mycoremediation (bioremediation with fungi) – growing mushrooms to clean the earth, Chemical Speciation and Bioavailability, 26: 3, 196-198,.

Cihangir N, Saglam N (1999) Removal of cadmium by Pleurotussajor-cajubasidiomycetes. ActaBiotechnol 19: 171–177.

Collins PJ, Dobson A (1997) Regulation of laccase gene transcription in *Trametes versicolor. Appl Environ Microbiol* 63: 3444–3450.

Da Luz JMR, Paes SA, Nunes MD, da Silva MCS, Kasuya MCM (2013) Degradation of Oxo-Biodegradable Plastic by *Pleurotus ostreatus. PLoS ONE* 8(8): 69386, doi: 10.1371/journal.pone.0069386.

Danýþ Ü (2010) Biosorption of copper(II) from aqueous solutions by Pleurotuscornucopiae.BALWOIS 2010, Ohrid, Republic of Macedonia, 25–29 May 2010.

Das N (2005) Heavy metals biosorption by mushrooms. *Indian J Natl Prod Resour* 4: 454–459.

Das N, Charumathi D, Vimala R (2007) Effect of pretreatment on Cd2+ biosorption by mycelia biomass of Pleurotus florida. *Afr J Biotechnol* 6: 2555–2558.

de Almeida LK, Burgess JE (2013) Biosorption and bioaccumulation of copper and lead by Phanerochaete and Pleurotus ostreatus. http: //www.ewisa. co.za/ literature/files/182_133 per cent 20Burgess.pdf. Accessed 20 June 2016.

Deng L, Zhang Y, Qin J, Wang X, Zhu X (2009) Biosorption of Cr(VI) from aqueous solutions by nonliving green algae Cladophoraalbida. *Miner Eng* 22: 372–377.

Drzewiecka K, Siwulski M, Mleczek M, Golinski P (2010) The Influence of elevated heavy metals content in substrate on morphology and physiology of King Oyster mushroom (Pleurotuseryngii) effects on human health. In: 15th International conference on heavy metals in the environment. http: //www. chem.pg.gda.pl/ichmet/.

Dulay RMR, De Castro MAEG, Coloma NB, Bernardo AP, Cruz AGD, Tiniola RC, Kalaw SP, Reyes RG (2015) Effects and myco-remediation of lead (Pb) in five Pleurotus mushrooms. *Int J Biol Pharm Allied Sci* 4(3): 1664–1677.

Eskander SB, Abd El-Aziz SM, El-Sayaad H, Saleh HM (2012) Cementation of bioproducts generated from biodegradation of radioactive cellulosic-based waste simulates by mushroom. ISRN Chemical Engineering, doi: 10.5402/2012/329676.

Favero N, Bressa G, Costa P (1990a) Response of Pleurotus ostreatusto cadmium exposure. Ecotoxicol Environ Safe 20(1): 1–6.

Favero N, Costa P, Massimino ML (1991) In vitro cadmium uptake by basidiomycetesPleurotus ostreatus.BiotechnolLett 13: 701–704.

Favero N, Costa P, Paolo Rocco G (1990b) Role of copper in cadmium metabolism in the basidiomycetes Pleurotus ostreatus. Comp BiochemPhysiol Part C Comp Pharmacol 97(2): 297–303.

Fawzy EM, Abdel-Motaal FF, EL-zayat SA (2017) Biosorption of heavy metals onto different eco-friendly substrates. J Toxicol Environ Health Sci 9(5): 35–44.

Firdousi SA (2017) Bioaccumulation and bio-absorptions of heavy metals by the mushroom from the soil. J Med Chem Drug Discov 2(3): 25–33.

Fontes Vieira PA, Gontijo DC, Vieira BC, Fontes EAF, Soares de Assunção L, Leite JPV, Oliveira MGdA, Kasuya MCM (2013) Antioxidant activities, total phenolics and metal contents in Pleurotus ostreatus mushrooms enriched with iron, zinc or lithium. LWT Food SciTechnol 54(2): 421–425.

Frutos I, García-Delgado C, Gárate A, Eymar E (2016) Biosorption of heavy metals by organic carbon from spent mushroom substrates and their raw materials. Int J Environ SciTechnol 13(11): 2713–2720.

Gabriel J, Capelari M, Rychlovský P, Krenželok M, Zadražil F (1996) Influence of cadmium on the growth of Agrocybe perfecta and two Pleurotus spp. and translocation from polluted substrate and soil to fruit bodies. Toxicol Environ Chem 56: 141–146.

Gaitán-Hernández R, Esqueda M, Gutiérrez A, Sánchez A, Beltrán-García M, Mata G (2006) Bioconversion of agrowastes by Lentinulaedodes: The high potential of viticulture residues. Appl Microbiol Biot 71: 432–439, doi: 10.1590/S1517-83822011000100004.

Gameiro PH, Nascimento JS, Rocha BH, Piana CF, Santos RA, Takahashi CS (2013) Antimutagenic effect of aqueous extract from Agaricusbrasiliensis on culture of human lymphocytes. J Med Food 16: 180–183, doi: 10.1089/jmf.2012.0068.

Gao Y, Dai X, Chen G, Ye J, Zhou S (2003) A randomized, placebo-controlled, multicenter study of Ganodermalucidum (W. Curt.: Fr.) Lloyd (Aphylloromycetidae) polysaccharides (Ganopoly R) in patients with advanced lung cancer. Int J Med Mushrooms 5: 369–381.

Gunatilake SK (2015) Methods of removing heavy metals from industrial wastewater. J MultidiscipEngSci Stud 1(1): 12–18.

Henini G, Laidani Y, FatihaSouahi F (2011) Study of adsorption of copper on biomass Pleurotusmutilusand the possibility of its regeneration by desorption. Energy Proced 6: 441–448.

Hlihor RM, Bulgariu L, Sobariu DL, Diaconu M, Tavares T, Gavrilescu M (2014) Recent advances in biosorption of heavy metals: support tools for biosorption equilibrium, kinetics and mechanism. Rev Roum Chim 59: 527–538.

Huo C-L, Shang Y-Y, Zheng J-J, He R-X, He XS (2011) The adsorption effect of three mushroom powder on Cu2+ of low concentration. In: International symposium on water resource and environmental protection, 20–22 May 2011. doi: 10.1109/ISWREP.2011.5893731.

Isikhuemhen, O.S. *et al.* (2011) The tropical white rot fungus, *Lentinus squarrosulus* Mont: lignocellolytic enzymes activities and sugar release from cornstalks under solid state fermentation. World J. Microbiol.Biotechnol.,28(5),1961-1966.

Ita BN, Ebong GA, Essien JP, Eduok SI (2008) Bioaccumulation potential of heavy metals in edible fungal sporocarps from the Niger delta region of Nigeria. Pak J Nutr 7: 93–97.

Ita BN, Essien JP, Ebong GA (2006) Heavy metal levels in fruiting bodies of edible and non-edible mushrooms from the Niger delta region of Nigeria. J Agric Soc Sci 2: 84–87.

Jain SK, Gujral GS, Jha NK, Vasudevan P (1988) Heavy metal uptake by Pleurotus sajor-caju from metal-enriched duckweed substrate. Biol Wastes 24: 275–282.

Jain SK, Gujral GS, Vasudevan P, Jha NK (1989) Uptake of heavy metals by Azollapinnata and their translocation onto the fruit bodies of Pleurotus sajor caju.J Ferment Bioeng 68(1): 64–67.

Jang KY,Cho SM, Seok SJ, Kong WS, Kim GH, Sung JM (2009) Screening of biodegradable function of indigenous ligno-degrading mushroom usingdyes. Mycobiology 37: 53–61, doi: 10.4489/MYCO.2009.37.1.053.

Javaid A, Bajwa R (2007) Biosorption of Cr(III) ions from tannery wastewater by Pleurotus ostreatus. Mycopathologia 5: 71–79.

Javaid A, Bajwa R (2008) Biosorption of electroplating heavy metals by some basiodiomycetes. Mycopathologia 6: 1–6.

Javaid A, Bajwa R, Shafique U, Anwar J (2011) Removal of heavy metals by adsorption on Pleurotus ostreatus. Biomass Bioenergy 35: 1675–1682.

Jibran AK, Milsee Mol JP (2011) Pleurotus sajor-caju Protein: A potential biosorptive agent. Adv Bio Tech 11: 25–27.

Jonathan SG, Fasidi IO, Ajayi AO, Adegeye O (2008) Biodegradation of Nigerian wood wastes by Pleurotus tuber-regium (Fries)Singer. Bioresource Technol 99: 807–811, doi: 10.1016/j.biortech.2007.01.005.

Kalac P, Svoboda L (2000) A review of trace element concentrations in edible mushrooms. Food Chem 69: 273–281.

Kamarudzaman AN, Tay CC, Amir A, Talib SA (2015) Biosorption of Mn(II) ions from aqueous solution by Pleurotusspent mushroom compost in a fixed-bed column. Proc Soc Behav Sci 195: 2709–2716.

Kang MY, Rico CW, Lee SC (2012) In vitro antioxidative and antimutagenic activities of oak mushroom (Lentinus edodes) and king Oyster mushroom (Pleurotus eryngii) byproducts. Food SciBiotechnol 21: 167–173, doi: 10.1007/s10068-012-0021-5.

Kapoor A, Viraraghavan T (1998) Biosorption of heavy metals on Aspergillus nigereffect of pretreatment. Bioresour Technol 63: 109–113.

Khitous M, Moussous S, Selatnia A, Kherat M (2015) Biosorption of Cd(II) by Pleurotus mutilus biomass in fixed-bed column: experimental and breakthrough curve analysis. Desalination Water Treat 57(35): 16559– 16570. doi: 10.1080/19443994.2015.1081625.

Kim HY, Yoon DH, Lee WH, Han SK, Shrestha B, Kim CH, Lim MH, Chang W, Lim S, Choi S, Song WO, Sung JM, Hwang KC, Kim TW (2007) Phellinus linteus inhibits inflammatory mediators by suppressing redox-based NF-jB and MAPKs activation in lipopolysaccharide-induced RAW 264.7 macrophage. J Ethno pharmacol 114: 307–315.

Kirk, T.K. and Farrell, R.L. (1987) Enzymatic combination– microbial degradation of lignin. Ann. Rev. Microbiol., 41, 465- 505.

Klein JM, Anziliero S, Camassola M, Grisa AMC, Brandalise RN, Zeni M (2012) Evaluation of metal biosorption by the fungus Pleurotus sajor-caju on modified polyethylene films. J Bioremed Biodeg 3: 152. doi: 10.4172/2155-6199.1000152: 5.

Kocaoba S, Arýsoy M (2011) The use of a white rot fungi (Pleurotus ostreatus) immobilized on Amberlite XAD-4 as a new biosorbent in trace metal determination. Bioresour Technol 102: 8035–8039.

Kodama N, Komuta K, Nanba H (2002) Can maitake MD-fraction aid cancer patients? Altern Med Rev 7: 236–239.

Kuforiji OO, Fasidi IO (2009) Biodegradation of agro-industrial wastes by an edible mushroom Pleurotus tuber-regium (Fr.). J Environ Biol 30: 355–358.

Kulshreshtha S, Mathur N, Bhatnagar P, Jain BL(2010) Bioremediation of industrial wastes through mushroom cultivation. J Environ Biol 31: 441–444.

Kulshreshtha S, Mathur N, Bhatnagar P(2014) Mushroom as a product and their role in mycoremediation. AMB Express 4: 29. doi: 10.1186/s13568-014-0029-8.

Lamar, R.T. and White, R.B. (2001) Mycoremediation: commercial status and recent developments. In: V.S. Magar *et al.* (eds), Proc. Sixth Int. Symp. on In Situ and on site Bioremediation, Vol. 6, pp. 263-278. San Diego, CA.

Lamrood PY, Ralegankar SD (2013) Biosorption of Cu, Zn, Fe, Cd, Pb and Ni by non treated biomass of some edible mushrooms. Asian J Exp Biol 4(2): 190–195.

Lang, E. Lang, G. E'Uer, I. Kleeberg, R. Martens, F. Zadrazil (1995)Interaction of white rot fungi and microorganisms leading to biodegradation of soil pollutants. In: W.J.Van de Brink *et al.* (eds), Proc. Fifth Int. FZK/TNo Conf. on contaminated soil, 30 Oct – 5 Nov 1995, Maustrient. The Netherlands.Contam. Soils, 95, 1277-1278.

Lechner BE, Papinutti VL (2006) Production of lignocellulosic enzymes during growth and fruiting of the edible fungus Lentinustigrinus on wheat straw. ProcessBiochem 41: 594–598, doi: 10.1016/j.procbio.2005.08.004.

Liew HH, Tay CC, Yong SK, Surif S, Abdul Talib S (2010) Biosorption characteristics of lead [Pb(II)] by Pleurotus ostreatusbiomass. In: Abstracts of the proceedings of international conference on science and social research (CSSR), Kuala Lumpur, 2010.

Luo D, Yf X, Tan ZL, Li XD (2013) Removal of Cu^{2+} ions from aqueous solution by the abandoned mushroom compost of Flammulinavelutipes. J Environ Biol 34: 359–365.

Maehara Y, Tsujitani S, Saeki H, Oki E, Yoshinaga K, Emi Y, Morita M, Kohnoe S, Kakeji Y, Yano T, Baba H (2012) Biological mechanism and clinical effect of protein-bound polysaccharide K, review of development and future perspectives. Surg Today 42: 8–28, doi: 10.1007/s00595-011-0075-7.

Majeed A, Jilani MI, Nadeem R, Hanif MA, Ansari TM (2012) Novel studies for the development of hybrid biosorbent. Int J ChemBiochemSci 2: 78–82.

Majeed A, Jilani MI, Nadeem R, Hanif MA, Ansari TM (2014) Adsorption of Pb(II) using novel Pleurotussajor-cajuand sunflower hybrid biosorbent. Environ ProtEng 40(2): 5–15.

Mandal TK, Baldrian P, Gabriel J, Nerud F, Zadraz/il F (1998) Effect of mercury on the growth of wood-rotting basidiomycetesPleurotus ostreatus, Pycnoporuscinnabarinus and Serpulalacrymans. Chemosphere 36(3): 435–440.

Manzi P, Aguzzi A, Pizzoferrato L (2001) Nutritional value of mushrooms widely consumed in Italy. Food Chem 73: 321–325.

Matsubara M, Lynch JM, AAM F, De Leij,(2006) A simple screening procedure for selecting fungi with potential for use in the bioremediation of contaminated land. Enzyme Microbial Technol., 39(7), 1365–1372.

Mosa KA, Saadoun I, Kumar K, Helmy M, Dhankher OP (2016) Potential biotechnological strategies for the cleanup of heavy metals and metalloids. Front Plant Sci 7: 1–14. doi: 10.3389/fpls.2016.00303.

Muszyñska B, D[1]browska M, Starek M, *et al.* 2019. Lentinula edodes Mycelium as Effective Agent for Piroxicam Mycoremediation. Front Microbiol. 2019;10: 313. Published 2019 Feb 21. doi: 10.3389/fmicb.2019.00313.

Nagy B, Mãicãneanu A, Indolean C, Mânzatu C, Silaghi-Dumitrescu MC (2013) Comparative study of Cd(II) biosorption on cultivated Agaricusbisporus and wild Lactariuspiperatus based biocomposites. Linear and nonlinear equilibrium modelling and kinetics J Taiwan InstChem E. doi: 10.1016/j.jtice.2013.08.013.

Nathan M. Watson (2020).Mycoremediation: Studies in Mycelium as an Agent of Bioremediation – Environmental ScienceBites Volume 2 by Brian H. Lower, Travis R. Shaul,Kylienne A. Shaul, and Ella M. published by Ohio State press.

Nnorom IC, Jarzyñska G, Falandysz J, Drewnowska M, Okoye I, Oji-Nnorom CG (2012) Occurrence and accumulation of mercury in two species of wild grown

Pleurotus mushrooms from southeastern Nigeria. Ecotoxicol Environ Safe 84: 78–83.

Olusola SA, Anslem EE (2010) Bioremediation of a crude oil polluted soil with PleurotusPulmonarius and GlomusMosseae using AmaranthusHybridus as a test plant. J BioremedBiodegrad 1: 111, doi: 10.4172/2155-6199.1000113.

Oyetayo VO, Adebayo AO, Ibileye A (2012) Assessment of the biosorption potential of heavy metals by Pleurotus tuber-regium. Int J Advanced Biol Res 2: 293–297.

Prakash V (2017) Mycoremediation of environmental pollutants. Int J Chem Tech Res 10(3): 149–155.

Puentes-Cárdenas IJ, Pedroza-Rodríguez AM, Navarrete-López M, Villegas-Garrido TL, Cristiani-Urbina E (2012) Biosorption of trivalent chromium from aqueous solutions by Pleurotus ostreatusbiomass. Environ Eng Manag J 11(10): 1741–1752.

Purkayastha RP, Mitra AK, Bhattacharyya B (1994) Uptake and toxicological effects of some heavy metals on Pleurotussajor-caju(Fr.) Singer. Ecotoxicol Environ Safe 27: 7–13.

Qazilbash AA (2004) Isolation and characterization of heavy metal tolerant biota from industrially polluted soils and their role in bioremediation. Biol Sci 41: 210–256.

Quarcoo A, Adotey G (2013) Determination of heavy metals in Pleurotus ostreatus(Oyster mushroom) and Termitomycesclypeatus(Termite mushroom) sold on selected markets in Accra, Ghana. Mycosphere 4(5): 960–967.

Radulescu C, Stihi C, Busuioc G, Gheboianu AI, Popescu IV (2010) Studies concerning heavy metals bioaccumulation of wild edible mushrooms from industrial area by using spectrometric techniques. Bull Environ ContamToxicol 84: 641–646.

Raj DD, Mohan B, Vidya Shetty BM (2011) Mushrooms in the remediation of heavy metals from soil. Int J Environ Pollut Control Manag 3(1): 89–101.

Rajput Y, Shit S, Shukla A, Shukla K (2011) Biodegradation of malachite green by wild mushroom of Chhatisgrah. J ExpSci 2: 69–72.

Rani P, Kalyani N, Prathiba K (2008) Evaluation of lignocellulosic wastes for production of edible mushrooms. Appl Biochem Biotechnol 151: 151–159, doi: 10.1007/s12010-008-8162-y.

Rhodes, C.J. (2012) Feeding and healing the world: through regenerative agriculture and permaculture. Sci. Prog., 95(4), 345-446.

Rhodes, C.J. (2013) Applications of bioremediation and phytoremediation. Sci. Prog., 96(4), 417-427.

Salman HA, Ibrahim MI, Tarek MM, Abbas HS (2014) Biosorption of heavy metals—a review. J Chem Sci Technol 3(4): 74–102.

Sing G, [a]engör S, Bhalla A, Kumar S, De J, Stewart B, Spycher N, M. Ginn T, Brent Peyton BM and Sani R(2014) Reoxidation of biogenic reduced uranium: a challenge toward bioremediation. Crit. Rev. Environ. Sci.Technol., 44(4), 391-415.

Singh J, Kant K, Sharma HB, Rana KS (2008) Bioaccumulation of cadmium in tissues of *Cirrihna mrigala* and *Catla catla*. Asian J ExpSci 22: 411–414.

Singh, H. (2006) Mycoremediation: fungal bioremediation. Wiley, Hoboken.

Soden DM, Dobson ADW (2001) Differential regulation of laccase gene expression in *Pleurotus sajor-caju*. Microbiology 147: 1755–1763.

Stamets, P. (2005). Mycelium running: how mushrooms can help save the world. Ten Speed Press, Berkley.

Suseem SR, Mary Saral A (2014) Biosorption of heavy metals using Pleurotuseous. J Chem Pharm Res 6(7): 2163–2168.

Sutherland C, Venkobachar C (2013) Equilibrium modeling of Cu (II) biosorption onto untreated and treated forest macro-fungus Fomesfasciatus. International Journal of Plant, Animal and Environment Sciences 3: 193–203.

Synytsya A, Mickova K, Synytsya A, Jablonsky I, Spevacek J, Erban V (2009) Glucans from fruit bodies of cultivated mushrooms *Pleurotus ostreatus* and *Pleurotus eryngii*: structure and potential prebiotic activity. CarbohydrPolym 76: 548–556. doi: 10.1016/j.carbpol.2008.11.02.

Tay CC, Liew HH, Abdul-Talib S, Redzwan G (2016) Bi-metal biosorption using Pleurotus ostreatusspent mushroom substrate (PSMS) as a biosorbent: isotherm, kinetic, thermodynamic studies and mechanism. Desalination Water Treat 57(20). http: //www.tandfonline.com/action/showCitFormats?.doi: http: //dx.doi.org/10.1080/19443994.2015. 1027957.

Tay CC, Liew HH, Yin C-Y, Abdul-Talib S, Surif S, Abdullah A, Yong SK (2011) Biosorptionofcadmium ions using Pleurotus ostreatus: growth kinetics, isotherm study and biosorption mechanism. Kor J ChemEng 28(3): 825–830.

Tay CC, Liew HH, Yin CY, Abdul-Talib S, Surif S, Suhaimi AA, Yong SK (2011) Biosorption of Cadmium ions using Pleurotus ostreatus: Growth kinetics, isotherm study and biosorption mechanism. Korean J ChemEng 28: 825–830, doi: 10.1007/s11814-010-0435-9.

Tay CC, Liew HH, Yong SK, Surif S, Abdul-Talib S (2009) Biosorption of lead(II) from aqueous solutions by Pleurotusas a toxicity biosorbent. In: Environmental science and technology conference (ESTEC2009), Kuala Terengganu Malaysia, Dec 7–8, 2009.

Tay CC, Redzwan G, Liew HH, Yong SK, Surif S, Abdul-Talib S (2012) Fundamental behavior for biosorption of divalencecations by Pleurotus mushroom spent-substrate. Malays J Sci 31: 40–44.

Tay CC, Redzwan G, Liew HH, Yong SK, Surif S, Abdul-Talib S Copper (II) (2010) Biosorption characteristic of Pleurotusspent mushroom compost. In: International conference on science and social research (CSSR 2010), Kuala Lumpur, Malaysia, Dec 5–7, 2010.

Tsujiyama S, Muraoka T, Takada N (2013) Biodegradation of 2,4-dichlorophenol by shiitake mushroom (*Lentinula edodes*) using vanillin as an activator. BiotechnolLett 35: 1079–1083, doi: 10.1007/s10529-013-1179-5.

Velásquez L, Dussan J (2009) Biosorption and bioaccumulation of heavy metals on dead and living biomass of Bacillus sphaericus. *J Hazard Mater* 167: 713–716. doi: 10.1016/j.jhazmat.2009.01.044.

Vimala R, Charumathi D, Nilanjana Das (2011) Packed bed column studies on Cd(II) removal from industrial wastewater by macrofungs *Pleurotus platypus.* Desalination 275: 291–296.

Vimala R, Das N (2011) Mechanism of Cd(II) adsorption by macrofungusPleurotus platypus. J Environ Sci 23: 288–293.

Wu M, Xu Y, Ding W, Li Y, Xu H (2016) Mycoremediation of manganese and phenanthrene by Pleurotuseryngiimycelium enhanced by tween 80 and saponin. Appl Microbiol Biotechnol 100: 7249–7261.

Xiangliang P, Jianlong W, Daoyong Z (2005) Biosorption of Pb(II) by Pleurotus ostreatusimmobilized in calcium alginate gel. Process Bio Chem40: 2799–2803.

Xiangliang P, Jianlong W, Daoyong Z (2009) Biosorption of Co(II) by immobilized Pleurotus ostreatus. Int J Environ Pollut 37: 289–298.

Xu F, Liu X, Chen Y, Zhang K, Xu H (2016) Self-assembly modified-mushroom nano composite for rapid removal of hexavalent chromium from aqueous solution with bubbling fluidized bed. Sci Rep 6. 26201. doi: 10.1038/srep26201. http: // www.nature.com/articles/srep26201.

Yalçinkaya Y, Arica MY, Soysal L, Bekta° S (2002) Cadmium and mercury uptake by immobilized Pleurotussapidus. Turk J Chem 26(3): 441–452 Yang T, Chen M-L, Wang J-H (2015) Genetic and chemical modification of cells for selective separation and analysis of heavy metals of biological or environmental significance. TrAC Trends Anal Chem 66: 90–102.

Yazdani M, Chee KY, Faridah A, Soon GT (2010) An in vitro study on the adsorption, absorption and uptake capacity of Zn by the bioremediatorTrichodermaatroviride. Environ Asia 3: 53–59.

Zhu FK, Qu L, Fan WX, Qiao MY, Hao HL, Wang XJ (2010) Assessment of heavy metals in some wild edible mushrooms collected from Yunnan Province, China. Environ Monit Assess 30: 61–62.

Chapter 14

Edible Flowers: An Extreme and Elegant Source of Minerals

Ankita Sharma[1] *and Reena Nair*[2]

[1]*Ph.D. Research Scholar,* [2]*Assistant Professor*
Department of Horticulture,
Jawaharlal Nehru Krishi Vishwa Vidhyalaya,
Jabalpur, Madhya Pradesh

Introduction and Ancient History of Edible Flowers

Edible flowers have a historic background since ancient Roman and Greek civilizations and also oriental countries like China are traditionally linked with these flowers in food preparation, as aroma enhancers, as flavouring intensifiers, and also for their aesthetic value (Gostin and Waisundara 2019; Vinokur *et al.*, 2006). Subsequently in the middle age, aromatic herbs and edible flowers were grown in the garden and were utilised in preparing infusions or syrups, confectionaries or jams, along with its use as food ingredients (diversified meals, from soups, to main courses or deserts) (Kaisoon *et al.*, 2011; Takahashi *et al.*, 2020).

Edible flowers are those flowers that can be utilized safely and are defined as innocuous non-toxic flowers consumed for health benefits (Alasalvar *et al.*, 2013). An appropriate identification of edible flowers is essential as some of them are poisonous.

With an increase in Globalisation, the pursuit of edible flowers is progressively increasing as it played a vital role in creating awareness among the consumers and also in the lifestyles. On a world-wide scale, people are demanding more appealing and appetizing food. The quality and aesthetic aspects have contributed towards the appearance of the meal consumed. The charisma and temptation of various dishes can be enhanced with the use of edible flowers. Edible flowers provide additional

Figure 14.1: Edible Flowers (Source: Fernandes *et al.*, 2018).

opportunities as it complements cut flowers or herb business and provide value-added products as it represents an important segment to expand the food market because of its suitable sensory and nutritional characteristics. According to Guine *et al.*, 2017 edible flowers are those that can be consumed by humans, as they have no contraindications and thus can be consumed safely because their components are absorbed when traveling through the intestine without causing toxicity. Keeping in view the socio-cultural factors, the study on edible flowers aims at popularizing and expanding this growing industry and encouraging local use of traditional flowers to preserve endangered traditions.

Over the last few years, the interest of researchers and consumers is increasing towards the use of edible flowers. Some of the flowers are entirely edible such as pumpkin flowers, in other cases, some specific portions of the flower such as petals and stigmas are used such as saffron for human consumption. Edible flowers are used in gastronomy among culinary chefs owing to their aesthetic as well as health properties. Some flowers are to be consumed in a small quantity as it also possesses some poisonous traits for instance linden flowers (*Tilia* spp.) which are only safe in small amount as consuming high quantity can cause damage to the heart. In an attempt to prevent possible complications along the digestive tract it is of paramount importance to consume EF only in small quantity and preferably without mixing different species when ingesting them for the first time. There are some flowers that can trigger allergic reactions such as the daisy that causes allergy in people suffering from asthma (Chen and Wei 2017a; Cunningham 2015; Mlcek and Rop 2011; Pires *et al.*, 2017).

Nutritional Composition

The nutritional composition is highly variable according to the type of flower, the most abundant component present in EF is water (variable from 70 to 95 per cent wet basis), carbohydrates (40–90 per cent dry basis), proteins and ash, being low in lipids. Additionally, they also constitutes micronutrients such as Vitamins A, C, B_2

Figurew 14.2: Flower Salad.
(*Source*: Travis Neighborward).

and B3 and dietary minerals such as K, P, Ca, Mg (Chen and Wei 2017b, Fernandes *et al.*, 2017, Navarro-González *et al.*, 2014; Petrova *et al.*, 2016a). Edible flowers also provide phyto-chemical compounds having pharmacological properties which are bioactive non-nutrient compounds such as polyphenols (Lu *et al.*, 2016; Fernandes *et al.*, 2017), essential amino acids, minerals (Drava *et al.*, 2020), polyunsaturated fatty acids, ascorbic acid, vitamin E (Miguel *et al.*, 2016; Pires *et al.*, 2017), and xantophylls (Fernandes *et al.*, 2018b) that can minimize the risk of major chronic diseases such as cancer, cardiovascular diseases and hepatic and gastro system. Some other properties possessed by EF are anti-inflammatory, diuretic, anthelmintic, modulatory of immune response, antimicrobial, and neuroprotective effects (Benvenuti *et al.*, 2016; Chen *et al.*, 2015; Kaisoon *et al.*, 2011; Loizzo *et al.*, 2016; Petrova *et al.*, 2016b). Edible flowers have a rich pigmentation which attracts pollinators.

Uses

The flowers have been used since ancient times to prepare tea, wine and supplements of butter, jams, marinades and sauces. They can be dried and used as medicines or herbs and can be crystallized to be used as desserts or accessories for desserts. The extracts of edible flowers or fresh flowers can be used as a viable alternative as natural antioxidant in food, pharmaceutical and cosmetology industries. Edible flowers are used in preparing culinary dishes such as roasts, stews, salads, soups and jams) as well as in liquid form (as alcoholic beverages and alcoholic or aqueous infusions) or even in the form of flavourings (as constituents of olive oils, other oils and vinegar.

Numerous flowers like Papaya (*Carica papaya*), Malabar nut (*Justicia adhatoda*), Midnight horror (*Oroxylum indicum*), Night jasmine (*Nyctanthus arbor tristis*) is consumed as curry with rice flour. (Deka *et al.*, 2014)

The flowers of Crape Jasmine (*Tabernaemontana divaricata*), camels foot tree (*Bauhinia spp.,*) and Agathi (*Sesbania grandiflora*) are eaten as Pokora.

Drum Stick Flower

Pond Weed Flower

Some flowers possess energy-giving constituents such as Flower of Drum stick (*Moringa oleifera*), Pumpkin (*Cucurbita moschata*), Arrow-leaved Oval-leafed pond weed (*Monochoria hastata*). A few of the flowers like common Arum (*Alocasia acuminate*) and Kanakachu (*Lasia spinosa*) are sources of calcium.

Flowers of water lily (*Nymphaea* spp.) is used after frying and is effective against anaemia.

Edible flowers such as Night Jasmine, *Phlogacanthus* spp. are bitter in taste and recommended to prevent digestive disorders like indigestion, loss of appetite, cold and cough, diabetes *etc.* (Deka *et al.*, 2014)

Figure 14.3: Banana Blossom (*Source*: Vegan Miam).

The floral spadix of banana (*Musa* spp.) raises the haemoglobin level in blood and dries up burnt portions in patients as it is a rich source of fibre and iron which assists in producing red blood cells. The tribes and communities of Assam also prepare a delicious delicacy with the flowers of banana and pigeon meat which is effective against low blood pressure, fever, cough *etc.* Banana blossoms contain calcium and potassium, along with vitamins A, C, and E. Intake of banana blossoms boosts milk production in lactating mothers. The soluble and insoluble fibres in the blossom are helpful for people suffering from irritable bowel syndrome. (Singh S., 2017). It is also rich in phytochemicals such as vitamins, flavonoids and proteins and has significant antibacterial and antioxidative properties. (Padma *et al.*, 2012)

Wild food plants can be used as natural colorant used to add color and fragrance to salad, soup, desserts and drinks and plays a vital role in livelihood of rural communities. (Rop *et al.*, 2012).

Edible flowers can also be consumed as dried, in mocktails, canned sugars are preserved in distillates. (Mlcek and Rop, 2011). *Calendula officinalis* gives the flavor of saffron due to orange-yellow coloured blossom. Carnation petals have a sweet-spicy flavour.

Blossoms are used to garnish dishes such as Nasturtium (*Tropaeolium majus*) having a peppy spicy flavour used in grilled vegetable platters and salads, Pansy (*Viola wittrockiana*) in soft cheeses and infusing color in syrups, star-shaped blossoms of borago (*Borago officinalis*) in fruit platter or as frozen ice cubes, rose hips (*Rose* spp.) in puddings and frozen treats and hibiscus (Hibiscus rosa-sinensis). Edible species of rose also exerted antiHIV-effects.

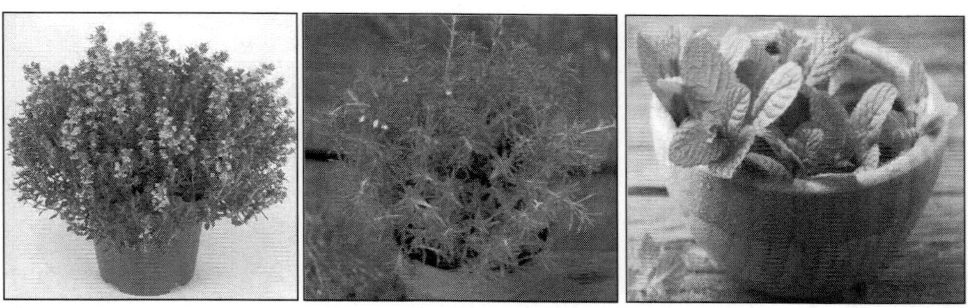

Thyme **Savory** **Mint**

Some herb flowers are edible which helps in improving the flavour of dishes, namely: alliums (leeks, chives, garlic), thyme (*Thymus vulgaris*), summer savory (*Satureja hortensis*), mint (*Mentha spp.*) and common sage (*Salvia officinalis*), which are used to improve the flavor of dishes. Floral blossoms of some fruit trees such as elderberry blossoms (*Sambucus spp*) and citrus blossoms (orange, lemon, lime, grapefruit, kumquat) can also be used in cuisine as syrups.

Chamomile (*Chamomilla recutita*) together with wild chrysanthemum and honeysuckle are also good sources of antinociceptive drugs. Dried blossoms of chamomile possess medicinal properties as it contain terpenoids and flavonoids. A herbal tea is prepared blended with medicinal herbs.

Chinese hibiscus *Hibiscus rosa-sinensis* has a mild tangy flavour can be brewed into teas and used in salads. It exerts anti-convulsive and contraceptive effects. It also promoted hair growth, wound healing and immunity.

Flowers of white chrysanthemum and day lily regulate sleep patterns and anti-depression. (Lin *et al.*, 2013).

Selection of Plant Material

Commonly cultivated annuals and perennials can be raised for their edible flowers. Edible flowers are not palatable and can be poisonous as well. The cultivars of the same species can differ in edibility. Popular edible flowers are chrysanthemum, daylily, lilac, mint, nasturtium, pansy, rose, tulip, and violet. Poisonous flowers from some common garden plants include anemone, autumn crocus, calla lily, daffodil, delphinium, foxglove, hyacinth, hydrangea, iris, lily of the valley and morning glory. Floral blossoms of vegetables such as pumpkins and fruits such as lemon, lime, orange can be used for culinary purpose.

Figure 14.4: Harvesting Edible Flower using Microtip Shears (Picture Courtesy: Garden Therapy).

Site Selection and Planting

The cultural practices are similar to those of ornamental flowers. Edible flowers prefer well-drained fertile soil and a sunny atmosphere. These flowers can even be maintained in raised beds to improve drainage and increase ease of harvest. Conducting a soil test to determine pH level of the soil is recommended before planting. Mulching can be effective to reduce weed pressure, maintain soil moisture and temperature, along with reducing soil splashing in heavy rains. Irrigation source, Trickle or drip irrigation is preferred to overhead irrigation so as to keep the foliage dry keeping to reduce the incidence of fungal and bacterial diseases. Annuals suffer losses due to frost. Growing the seedlings in nursery and then transplanting in the main field flowers soon and yields more as compared to direct-seeded crop can be adopted by farmers to capture the early market and then direct-seed later plantings. Transplanting can be done in a greenhouse or cold frame. Perennials, such as daylily, should be planted in late summer or fall for the best yield of blooms.

Insect Pests and Disease Management

Buyers and consumers would prefer edible flowers free from insect pests and diseases. Insect management is done using cultural and biological practices as use of chemical pesticides and insecticides would be harmful. Insect control is done by hand-picking harmful insects and using beneficial insects to help decrease insect populations. Following better management and cultural practices and diversifying plantings aids in controlling both insect and disease problems. Weeds should be controlled as it competes with growing flowers and reduces plant quality and quantity, but also raises labor costs by increasing the time required for harvest. Flowers which are to be grown for human consumption should be free from pesticides or other chemicals thus providing organic growers a production edge. (University of Kentucky School of Human Environmental Sciences, 1997).

Harvesting of Edible Flowers

Edible flowers should be harvested during late evening or early morning at the time of cool part of the day when the flower is at full bloom as they are in the best flavour and have the highest water content so remain fresh and perky. Ideally, edible flowers should be cut just before use of should be processed on the day of harvest itself. Use a clean pair of micro-tip pruning shears to cut the bloom just below the floral head. Flowers free from bruises, marks diseases or attacks of insects should be chosen for marketing and use. Wilted or faded flowers are avoided due to their bitter and unappealing flavour. Unopened blossoms or those past their prime are not selected for harvest. (Newman and O'Connor, 2013). Flowers when collected for eating purposes should not be sprayed with pesticides or untreated manure because they can be toxic or cause illness due to the presence of pathogens. Some floral parts such as the pedicel, sepals, pistils and stamens of most flowers are removed prior to use. Pollens of some flowers can detract flower's flavour and cause hay fever, asthma, allergies. The sepals are sour except that of Violas, Pancies and hence should be removed. In Flowers of rose, lavender, tulip, calendula, and chrysanthemum only the petals are edible. The flowers should be cooled immediately after harvest to maintain freshness. Flowers with white and bitter petal bases such as chrysanthemum, dianthus, marigold, and rose should be removed as it is bitter to taste.

Labour Requirements

The production of the edible flowers is a labor and management intensive practice. All packages and practices like Planting, weeding, and harvesting require trained labour. Extra time and care are taken to transport the edible flower product from the field to the market. Additional packing labour is required to pack different edible flower varieties and/or colours.

Economic Considerations

A higher profit can be obtained by producing edible flowers as the demand is increasing due to its nutritional components. A wide difference is observed in species, handling, marketing, and delivery. The producers can receive a reasonable return from the land and management. Higher selling prices can compensate a producer for their harvest and handling labour. Chefs and consumers prefer purchasing a well-packaged product. The addition of other edible varieties to a mix, especially those with desirable fragrance, enhanced consumer interest.

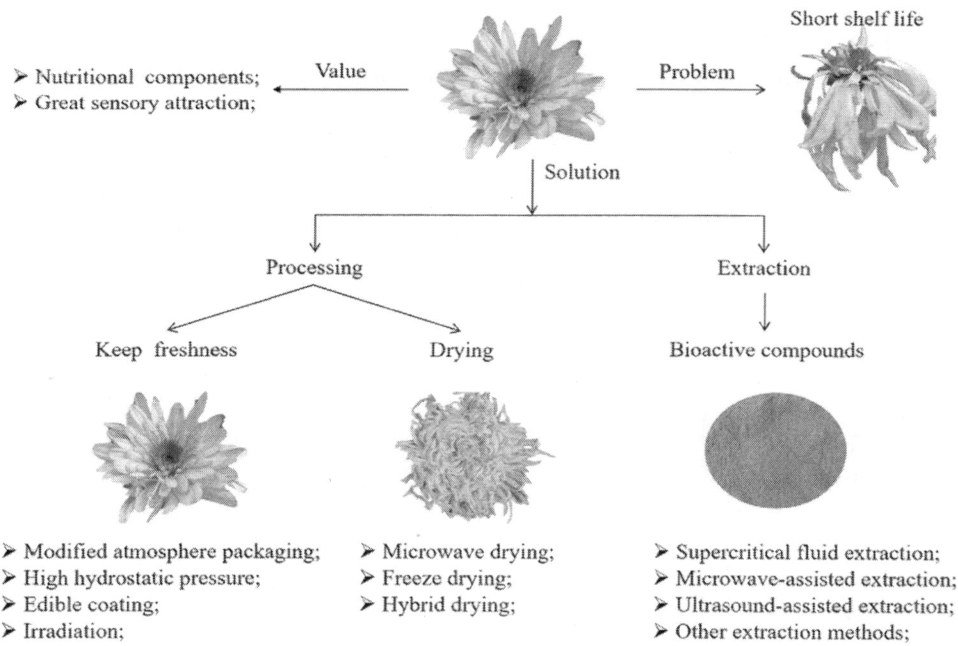

Figure 14.5: Processing Techniques to Enhance Shelf Life of Edible Flowers. (*Source*: Zhao *et al.,* 2019).

Postharvest Technologies

The Edible flowers are extremely perishable and delicate. Edible flowers have a great sensory attraction. They have an extremely short shelf life which limits their commercial use but with the use of novel processing tools including modified atmospheric packaging, high hydrostatic pressure, irradiation and edible coating can keep flower fresh. Drying techniques including microwave drying, freeze-drying and hybrid drying maintains an optimal state of floral material. Extraction methods and microencapsulation can be used to extract bioactive compounds (Zhao *et al.*, 2019).

Postharvest, the flowers are susceptible to petal abscission, discolouration, wilting, dehydration and tissue browning so refrigeration of edible flowers before packaging for small trips, by keeping them in cold storage between 4 and 6 °C, can increase their durability from 2 to 5 days after harvest (Kou *et al.*, 2012). To prevent desiccation and to preserve their frail structure, edible flowers should be typically packaged in small, rigid, plastic (or plastic-wrapped) packages. Flowers are safe and edible, including the flowers of most culinary herbs, proper identification is essential. Flowers sold by florist or in garden centers are not labelled for edible purpose since they have been treated with pesticides. Flowers should be dried and used immediately after harvest to produce value-added flower products.

Marketing of Edible Flowers

Flowers can be marketed fresh, dried, candied or in prepackaged salads. Packaging different coloured flowers and variety of edible flowers in the same container are appealing to customers and have a positive reaction. Value-added products having feature edible flowers offer additional marketing opportunities. Minced flowers make a colourful and flavourful addition to herbal butters, cheese spreads, jellies, and jams. Dried flowers products used in teas, to add flavor to wines, pot pourri, wreath, paintedgourd or for processing like rose water gulkand, perfumes, essential oils, insect repellents *etc.* The value addition is acceptable for domestic and export market and ensures high premium to producer. Value addition helps to avoid wastage during glut period and farmers can earn better remuneration. It also helps in employment generation at producer, whole seller and at industrial level.

Conclusion

Edible flowers are innocuous flowers with health benefits which not only add colours to our plate but also play a vital role in the livelihoods of rural communities. There are underutilized plant species that possess multiple uses. Flowers add great visual to dishes and are valuable for gastronomy. Edible flowers are potential food ingredients and avoid potential hazards. The demands of flowers in market are increasing which is a positive step towards food security and management of plant resources ethically. These flowers are a new and promising foodstuff species as have a high nutritional value, antioxidant capacity and attractive appearance. Postharvest strategies can be tested to maintain the nutraceautical properties of edible flowers.

REFERENCES

Alasalvar C, Pelvan E, Özdemir KS, Kocadagh T. Mogol BA, Pash AA., Özcan N, Özçelik, B, Gökmen V 2013, Compositional, nutritional, and functional characteristics of instant teas produced from low- and high-quality black teas. *J. Agric. Food Chem.* 61, 7529–7536.

Benvenuti S, Bortolotti E, Maggini R. 2016. Antioxidant power, anthocyanin content and organoleptic performance of edible flowers. *Sci Horticulturae.* 199: 170–7.

Chen N-H, Wei S. 2017. Factors influencing consumers' attitudes towards the consumption of edible flowers. *Food Qual Prefer.* 56: 93–100.

Cunningham E. 2015.What nutritional contribution do edible flowers make? J Acad *Nutr Dietetics,* 115: 856.

Deka k, Nath N, 2014. Documentation of edible flowers of western Assam. American *Journal of Phytomedicine and Clinical Therapeutics;* 2 (10), 1124-1140

Drava G, Iobbi V, Govaerts R, Minganti V, Copetta A, Ruffoni B 2020, Trace elements in edible flowers from Italy: further insights into health benefits and risks to consumers. *Molecules,* 25: 2891.

Edible Flowers University of Kentucky School of Human Environmental Sciences, 1997.

Edible Flowers Colorado State University, 2009.

Fernandes L, Casal S, Pereira JA, Saraiva JA, Ramalhosa E. Edible flowers 2017, A review of the nutritional, antioxidant, antimicrobial properties and effects on human health. *J Food Comp Anal.* 60: 38–50.

Fernandes L, Ramalhosa E, Pereira JA, Saraiva JA and Casal S. 2018b. The unexplored potential of edible flowers lipids. *Agriculture.* 8, 1–23.

Guiné R, Santos E, Correia P. 2017. Edible flowers: Knowledge and consumption habits. *Acta Sci Nut Health* 1: 18–22.

Gostin A-I, Waisundara VY. 2019. Edible flowers as functional food: a review on artichoke (*Cynara cardunculus* L.). *Trends Food Sci Technol.* 86: 381–91.

Kaisoon O, Siriamornpun S, Weerapreeyakul N, Meeso N. 2011.Phenolic compounds and antioxidant activities of edible flowers from Thailand. *J Funct Foods,* 3: 88–99.

Kou, Liping and Turner, Ellen and Luo, Yaguang. 2012. Extending the Shelf Life of Edible Flowers with Controlled Release of 1-Methylcyclopropene and Modified Atmosphere Packaging. *Journal of Food Science.* 77. S188-93.

Loizzo MR, Pugliese A, Bonesi M, Tenuta MC, Menichini F, Xiao J, 2016. Edible flowers: A rich source of phytochemicals with antioxidant and hypoglycemic properties. *J Agric Food Chem.* 64: 2467–74.

Lu B, Li M and Yin R. 2016. Phytochemical content, health benefits, and toxicology of common edible flowers: a review (2000–2015). *Crit. Rev. Food Sci. Nutr.* 56 (Suppl. 1); S130–S148.

Miguel M, Barros L, Pereira C, Calhelha RC, Garcia PA, Castro MÁ 2016. Chemical characterization and bioactive properties of two aromatic plants: *Calendula officinalis* L. (flowers) and *Mentha cervina* L. (leaves). *Food Funct.* 7, 2223–2232.

Mlcek J and Rop O. 2011. Fresh edible flowers of ornamental plants are a new source of nutraceutical foods. *Trends Food Sci. Technol.,* 22: 561–569.

Navarro-González I, González-Barrio R, García-Valverde V, Bautista-Ortín AB, Periago MJ. 2014. Nutritional composition and antioxidant capacity in edible flowers: Characterisation of phenolic compounds by HPLC-DAD-ESI/MSn. *Int J Mol Sci.,* 16: 805–22.

Padma BS, Tin HS, Chye FY, Abdullah MI. 2012, Antibacterial and antioxidative activities of the various solvent extracts of banana (*Musa paradisiaca* cv. Mysore) inflorescences. *Biological Sciences.* 12(2): 62-73.

Petrova I, Petkova N, Ivanov I. 2016b. Five edible flowers–valuable source of antioxidants in human nutrition. *Int J Pharmacog Phytochem Res.*8: 604–10.

Pires TCSP, Dias MI, Barros L, Ferreira ICFR. 2017. Nutritional and chemical characterization of edible petals and corresponding infusions: valorization as new food ingredients. *Food Chem.* 220: 337–43.

Newman SE Colorado State University Extension greenhouse crops specialist and professor, horticulture and landscape architechture; and A. Stoven O'Connor, Extension horticulture agent, Larimer County. 12/96. Revised 10/20.

Singh S. 2017. Banana blossom-an understated food with high functional benefits. *International Journal of Current Research.* 9(01): 44516-44519.

Takahashi JA, Rezende FAGG, Moura MAF, Dominguete LCB, Sande D. 2020. Edible flowers: Bioactive profile and its potential to be used in food development. *Food Res Int.* 129: 108868.

Vinokur Y, Rodov V, Reznick N, Goldman G, Horev B, Umiel N2006. Rose petal tea as an antioxidant-rich beverage: *Cultivar Effects J. Food Sci.*71: S42–7.

Zhao L, Fan H, Zhang M, Chitrakar B, Bhandari B, Wang B. 2019. Edible flowers: Review of flower processing and extraction of bioactive compounds by novel technologies, *Food Research International.* Volume 126; 9963-9969.

Index

K

L

M

N

O

P